ELECTRIC POWER DISTRIBUTION EQUIPMENT AND SYSTEMS

ELECTRIC POWER DISTRIBUTION EQUIPMENT AND SYSTEMS

T. A. Short
EPRI Solutions, Inc.
Schenectady, NY

CRC Press
Taylor & Francis Group
Boca Raton London New York

CRC Press is an imprint of the
Taylor & Francis Group, an **informa** business
A TAYLOR & FRANCIS BOOK

The material was previously published in *Electric Power Distribution Handbook* © CRC Press LLC 2004.

CRC Press
Taylor & Francis Group
6000 Broken Sound Parkway NW, Suite 300
Boca Raton, FL 33487-2742

First issued in paperback 2019

© 2006 by Taylor & Francis Group, LLC
CRC Press is an imprint of Taylor & Francis Group, an Informa business

No claim to original U.S. Government works

ISBN-13: 978-0-8493-9576-5 (hbk)
ISBN-13: 978-0-367-39167-6 (pbk)
Library of Congress Card Number 2005052135

Library of Congress Cataloging-in-Publication Data

Short, T.A. (Tom A.), 1966-
   Electric power distribution equipment and systems / Thomas Allen Short.
      p. cm.
   Includes bibliographical references and index.
   ISBN 0-8493-9576-3 (alk. paper)
   1. Electric power distribution--Equipment and supplies. I. Title.

TK3091.S466 2005
621.319--dc22   2005052135

Visit the Taylor & Francis Web site at
http://www.taylorandfrancis.com

and the CRC Press Web site at
http://www.crcpress.com

Dedication

To the future. To Jared. To Logan.

Preface

In industrialized countries, distribution systems deliver electricity literally everywhere, taking power generated at many locations and delivering it to end users. Generation, transmission, and distribution—of the big three components of the electricity infrastructure, the distribution system gets the least attention. Yet, it is often the most critical component in terms of its effect on reliability and quality of service, cost of electricity, and aesthetic (mainly visual) impacts on society.

Like much of the electric utility industry, several political, economic, and technical changes are pressuring the way distribution systems are built and operated. Deregulation has increased pressures on electric power utilities to cut costs and has focused emphasis on reliability and quality of electric service. The great fear of deregulation is that service will suffer because of cost cutting. Regulators and utility consumers are paying considerable attention to reliability and quality. Customers are pressing for lower costs and better reliability and power quality. The performance of the distribution system determines greater than 90% of the reliability of service to customers (the high-voltage transmission and generation system determines the rest). If performance is increased, it will have to be done on the distribution system. Utilities are looking for the most cost-effective and efficient management of their distribution assets.

This book is a spinoff from the *Electric Power Distribution Handbook* (2004) that includes the portions of that handbook that target equipment and applications of equipment. It includes overhead designs, underground issues and applications, and voltage regulation and capacitor applications. Managing these assets is key to controlling costs, regulating voltage, controlling maintenance, and managing failures. Proper specification, application, and maintenance will improve equipment reliability, which will help reduce costs, improve safety, and improve customer reliability.

I hope you find useful information in this book. If it's not in here, hopefully, one of the many bibliographic references will lead you to what you're looking for. Please feel free to e-mail me feedback on this book including errors, comments, opinions, or new sources of information—I'd like to hear from you. Also, if you need my help with any interesting consulting or research opportunities, I'd love to hear from you.

Tom Short
EPRI Solutions, Inc.
Schenectady, NY
t.short@ieee.org

Acknowledgments

First and foremost, I'd like to thank my wife Kristin—thank you for your strength, thank you for your help, thank you for your patience, and thank you for your love. My play buddies, Logan and Jared, energized me and made me laugh. My family was a source of inspiration. I'd like to thank my parents, Bob and Sandy, for their influence and education over the years.

EPRI Solutions, Inc. (formerly EPRI PEAC) provided a great deal of support on this project. I'd like to recognize the reviews, ideas, and support of Phil Barker and Dave Crudele here in Schenectady, New York, and also Arshad Mansoor, Mike Howard, Charles Perry, Arindam Maitra, and the rest of the energetic crew in Knoxville, Tennessee.

Many other people reviewed portions of the draft and provided input and suggestions including Dave Smith (Power Technologies, Inc.), Dan Ward (Dominion Virginia Power), Jim Stewart (Consultant, Scotia, NY), Conrad St. Pierre (Electric Power Consultants), Karl Fender (Cooper Power Systems), John Leach (Hi-Tech Fuses, Inc.), and Rusty Bascom (Power Delivery Consultants, LLC).

Thanks to Power Technologies, Inc. for opportunities and mentoring during my early career with the help of several talented, helpful engineers, including Jim Burke, Phil Barker, Dave Smith, Jim Stewart, and John Anderson. Over the years, several clients have also educated me in many ways; two that stand out include Ron Ammon (Keyspan, retired) and Clay Burns (National Grid).

EPRI has been supportive of this project, including a review by Luther Dow. EPRI has also sponsored a number of interesting distribution research projects that I've been fortunate enough to be involved with, and EPRI has allowed me to share some of those efforts here.

As a side-note, I'd like to recognize the efforts of linemen in the electric power industry. These folks do the real work of building the lines and keeping the power on. As a tribute to them, a trailer at the end of each chapter reveals a bit of the lineman's character and point of view.

About the Author

Mr. Short has spent most of his career working on projects helping utilities improve their reliability and power quality. He performed lightning protection, reliability, and power quality studies for many utility distribution systems while at Power Technologies, Inc. from 1990 through 2000. He has done extensive digital simulations of T&D systems using various software tools including EMTP to model lightning surges on overhead lines and underground cables, distributed generators, ferroresonance, faults and voltage sags, and capacitor switching. Since joining EPRI PEAC in 2000 (now EPRI Solutions, Inc.), Mr. Short has led a variety of distribution research projects for EPRI, including a capacitor reliability initiative, a power quality handbook for distribution companies, a distributed generation workbook, and a series of projects directed at improving distribution reliability and power quality.

As chair of the IEEE Working Group on the Lightning Performance of Distribution Lines, he led the development of IEEE Std. 1410-1997, *Improving the Lightning Performance of Electric Power Overhead Distribution Lines*. He was awarded the 2002 Technical Committee Distinguished Service Award by the IEEE Power Engineering Society for this effort.

Mr. Short has also performed a variety of other studies including railroad impacts on a utility (flicker, unbalance and harmonics), load flow analysis, capacitor application, loss evaluation, and conductor burndown. Mr. Short has taught courses on reliability, power quality, lightning protection, overcurrent protection, harmonics, voltage regulation, capacitor application, and distribution planning.

Mr. Short developed the Rpad engineering analysis interface (www.Rpad.org) that EPRI Solutions, Inc. is using to offer engineering, information, mapping, and database solutions to electric utilities. Rpad is an interactive, web-based analysis program. Rpad pages are interactive workbook-type sheets based on R, an open-source implementation of the S language (used to make many of the graphs in this book). Rpad is an analysis package, a web-page designer, and a gui designer all wrapped in one. Rpad makes it easy to develop powerful data-analysis applications that can be easily shared on a company intranet.

Mr. Short graduated with a master's degree in electrical engineering from Montana State University in 1990 after receiving a bachelor's degree in 1988.

Contents

Credits

Tables 4.3 to 4.7 and 4.13 are reprinted with permission from IEEE Std. C57.12.00-2000. *IEEE Standard General Requirements for Liquid-Immersed Distribution, Power, and Regulating Transformers.* Copyright 2000 by IEEE.

Figure 4.17 is reprinted with permission from ANSI/IEEE Std. C57.105-1978. *IEEE Guide for Application of Transformer Connections in Three-Phase Distribution Systems.* Copyright 1978 by IEEE.

Tables 6.2, 6.4, and 6.5 are reprinted with permission from IEEE Std. 18-2002. *IEEE Standard for Shunt Power Capacitors.* Copyright 2002 by IEEE.

Table 6.3 is reprinted with permission from ANSI/IEEE Std. 18-1992. *IEEE Standard for Shunt Power Capacitors.* Copyright 1993 by IEEE.

1

Fundamentals of Distribution Systems

Electrification in the early 20th century dramatically improved productivity and increased the well-being of the industrialized world. No longer a luxury — now a necessity — electricity powers the machinery, the computers, the health-care systems, and the entertainment of modern society. Given its benefits, electricity is inexpensive, and its price continues to slowly decline (after adjusting for inflation — see Figure 1.1).

Electric power distribution is the portion of the power delivery infrastructure that takes the electricity from the highly meshed, high-voltage transmission circuits and delivers it to customers. Primary distribution lines are "medium-voltage" circuits, normally thought of as 600 V to 35 kV. At a distribution substation, a substation transformer takes the incoming transmission-level voltage (35 to 230 kV) and steps it down to several distribution primary circuits, which fan out from the substation. Close to each end user, a distribution transformer takes the primary-distribution voltage and steps it down to a low-voltage secondary circuit (commonly 120/240 V; other utilization voltages are used as well). From the distribution transformer, the secondary distribution circuits connect to the end user where the connection is made at the service entrance. Figure 1.2 shows an overview of the power generation and delivery infrastructure and where distribution fits in. Functionally, distribution circuits are those that feed customers (this is how the term is used in this book, regardless of voltage or configuration). Some also think of distribution as anything that is radial or anything that is below 35 kV.

The distribution infrastructure is extensive; after all, electricity has to be delivered to customers concentrated in cities, customers in the suburbs, and customers in very remote regions; few places in the industrialized world do not have electricity from a distribution system readily available. Distribution circuits are found along most secondary roads and streets. Urban construction is mainly underground; rural construction is mainly overhead. Suburban structures are a mix, with a good deal of new construction going underground.

A mainly urban utility may have less than 50 ft of distribution circuit for each customer. A rural utility can have over 300 ft of primary circuit per customer.

Several entities may own distribution systems: municipal governments, state agencies, federal agencies, rural cooperatives, or investor-owned utili-

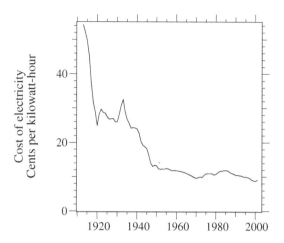

FIGURE 1.1
Cost of U.S. electricity adjusted for inflation to year 2000 U.S. dollars. (Data from U.S. city average electricity costs from the U.S. Bureau of Labor Statistics.)

ties. In addition, large industrial facilities often need their own distribution systems. While there are some differences in approaches by each of these types of entities, the engineering issues are similar for all.

For all of the action regarding deregulation, the distribution infrastructure remains a natural monopoly. As with water delivery or sewers or other utilities, it is difficult to imagine duplicating systems to provide true competition, so it will likely remain highly regulated.

Because of the extensive infrastructure, distribution systems are capital-intensive businesses. An Electric Power Research Institute (EPRI) survey found that the distribution plant asset carrying cost averages 49.5% of the total distribution resource (EPRI TR-109178, 1998). The next largest component is labor at 21.8%, followed by materials at 12.9%. Utility annual distribution budgets average about 10% of the capital investment in the distribution system. On a kilowatt-hour basis, utility distribution budgets average 0.89 cents per kilowatt-hour (see Table 1.1 for budgets shown relative to other benchmarks).

Low cost, simplification, and standardization are all important design characteristics of distribution systems. Few components and/or installations are individually engineered on a distribution circuit. Standardized equipment and standardized designs are used wherever possible. "Cookbook" engineering methods are used for much of distribution planning, design, and operations.

Distribution planning is the study of future power delivery needs. Planning goals are to provide service at low cost and high reliability. Planning requires a mix of geographic, engineering, and economic analysis skills. New circuits (or other solutions) must be integrated into the existing distribution system within a variety of economic, political, environmental, electrical, and geographic constraints. The planner needs estimates of load

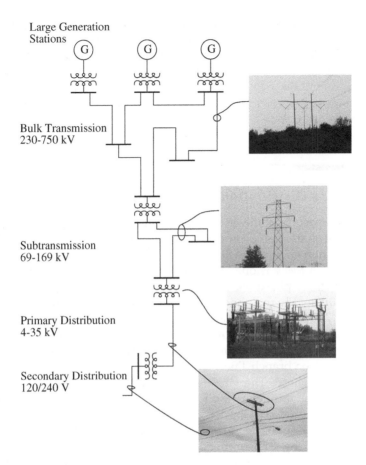

FIGURE 1.2
Overview of the electricity infrastructure.

TABLE 1.1

Surveyed Annual Utility Distribution Budgets in
U.S. Dollars

	Average	Range
Per dollar of distribution asset	0.098	0.0916–0.15
Per customer	195	147–237
Per thousand kWH	8.9	3.9–14.1
Per mile of circuit	9,400	4,800–15,200
Per substation	880,000	620,000–1,250,000

Source: EPRI TR-109178, *Distribution Cost Structure — Methodology and Generic Data*, Electric Power Research Institute, Palo Alto, CA, 1998.

growth, knowledge of when and where development is occurring, and local development regulations and procedures. While this book has some material that should help distribution planners, many of the tasks of a planner, like load forecasting, are not discussed. For more information on distribution planning, see Willis's *Power Distribution Planning Reference Book* (1997), IEEE's *Power Distribution Planning* tutorial (1992), and the *CEA Distribution Planner's Manual* (1982).

1.1   Primary Distribution Configurations

Distribution circuits come in many different configurations and circuit lengths. Most share many common characteristics. Figure 1.3 shows a "typical" distribution circuit, and Table 1.2 shows typical parameters of a distribution circuit. A *feeder* is one of the circuits out of the substation. The main feeder is the three-phase backbone of the circuit, which is often called the *mains* or *mainline*. The mainline is normally a modestly large conductor such as a 500- or 750-kcmil aluminum conductor. Utilities often design the main feeder for 400 A and often allow an emergency rating of 600 A. Branching from the mains are one or more *laterals*, which are also called taps, lateral taps, branches, or branch lines. These laterals may be single-phase, two-phase, or three-phase. The laterals normally have fuses to separate them from the mainline if they are faulted.

The most common distribution primaries are four-wire, multigrounded systems: three-phase conductors plus a multigrounded neutral. Single-phase loads are served by transformers connected between one phase and the neutral. The neutral acts as a return conductor and as an equipment safety ground (it is grounded periodically and at all equipment). A single-phase line has one phase conductor and the neutral, and a two-phase line has two phases and the neutral. Some distribution primaries are three-wire systems (with no neutral). On these, single-phase loads are connected phase to phase, and single-phase lines have two of the three phases.

There are several configurations of distribution systems. Most distribution circuits are radial (both primary and secondary). Radial circuits have many advantages over networked circuits including

- Easier fault current protection
- Lower fault currents over most of the circuit
- Easier voltage control
- Easier prediction and control of power flows
- Lower cost

Distribution primary systems come in a variety of shapes and sizes (Figure 1.4). Arrangements depend on street layouts, the shape of the area covered

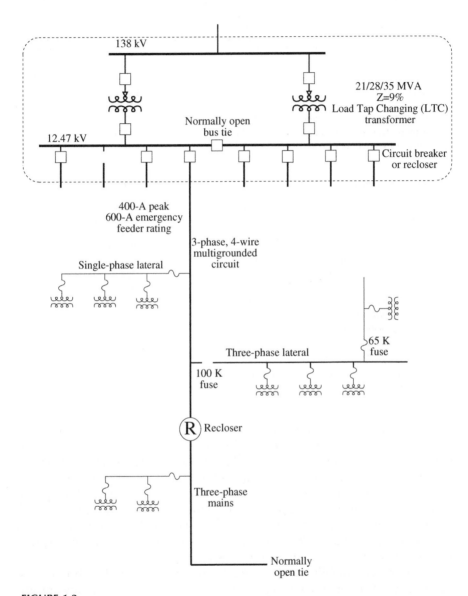

FIGURE 1.3
Typical distribution substation with one of several feeders shown (many lateral taps are left off). (Copyright © 2000. Electric Power Research Institute. 1000419. *Engineering Guide for Integration of Distributed Generation and Storage Into Power Distribution Systems*. Reprinted with permission.)

TABLE 1.2

Typical Distribution Circuit Parameters

	Most Common Value	Other Common Values
Substation characteristics		
Voltage	12.47 kV	4.16, 4.8, 13.2, 13.8, 24.94, 34.5 kV
Number of station transformers	2	1–6
Substation transformer size	21 MVA	5–60 MVA
Number of feeders per bus	4	1–8
Feeder characteristics		
Peak current	400 A	100–600 A
Peak load	7 MVA	1–15 MVA
Power factor	0.98 lagging	0.8 lagging–0.95 leading
Number of customers	400	50–5000
Length of feeder mains	4 mi	2–15 mi
Length including laterals	8 mi	4–25 mi
Area covered	25 mi^2	0.5–500 mi^2
Mains wire size	500 kcmil	4/0–795 kcmil
Lateral tap wire size	1/0	#4–2/0
Lateral tap peak current	25 A	5–50 A
Lateral tap length	0.5 mi	0.2–5 mi
Distribution transformer size (1 ph)	25 kVA	10–150 kVA

by the circuit, obstacles (like lakes), and where the big loads are. A common suburban layout has the main feeder along a street with laterals tapped down side streets or into developments. Radial distribution feeders may also have extensive branching — whatever it takes to get to the loads. An *express feeder* serves load concentrations some distance from the substation. A three-phase mainline runs a distance before tapping loads off to customers. With many circuits coming from one substation, a number of the circuits may have express feeders; some feeders cover areas close to the substation, and express feeders serve areas farther from the substation.

For improved reliability, radial circuits are often provided with normally open tie points to other circuits as shown in Figure 1.5. The circuits are still operated radially, but if a fault occurs on one of the circuits, the tie switches allow some portion of the faulted circuit to be restored quickly. Normally, these switches are manually operated, but some utilities use automated switches or reclosers to perform these operations automatically.

A primary-loop scheme is an even more reliable service that is sometimes offered for critical loads such as hospitals. Figure 1.6 shows an example of a primary loop. The key feature is that the circuit is "routed through" each

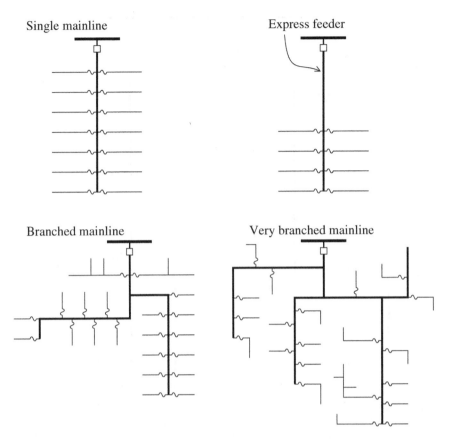

FIGURE 1.4
Common distribution primary arrangements.

critical customer transformer. If any part of the primary circuit is faulted, all critical customers can still be fed by reconfiguring the transformer switches.

Primary-loop systems are sometimes used on distribution systems for areas needing high reliability (meaning limited long-duration interruptions). In the open-loop design where the loop is left normally open at some point, primary-loop systems have almost no benefits for momentary interruptions or voltage sags. They are rarely operated in a closed loop. A widely reported installation of a sophisticated *closed* system has been installed in Orlando, FL, by Florida Power Corporation (Pagel, 2000). An example of this type of closed-loop primary system is shown in Figure 1.7. Faults on any of the cables in the loop are cleared in less than six cycles, which reduces the duration of the voltage sag during the fault (enough to help many computers). Advanced relaying similar to transmission-line protection is necessary to coordinate the protection and operation of the switchgear in the looped system. The relaying scheme uses a transfer trip with permissive over-reaching (the relays at each end of the cable must agree there is a fault

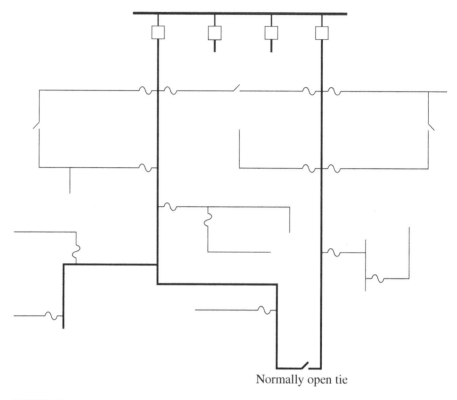

Normally open tie

FIGURE 1.5
Two radial circuits with normally open ties to each other. (Copyright © 2000. Electric Power Research Institute. 1000419. *Engineering Guide for Integration of Distributed Generation and Storage Into Power Distribution Systems.* Reprinted with permission.)

between them with communications done on fiberoptic lines). A backup scheme uses directional relays, which will trip for a fault in a certain direction unless a blocking signal is received from the remote end (again over the fiberoptic lines).

Critical customers have two more choices for more reliable service where two primary feeds are available. Primary selective and secondary selective schemes both are normally fed from one circuit (see Figure 1.8). So, the circuits are still radial. In the event of a fault on the primary circuit, the service is switched to the backup circuit. In the primary selective scheme, the switching occurs on the primary, and in the secondary selective scheme, the switching occurs on the secondary. The switching can be done manually or automatically, and there are even static transfer switches that can switch in less than a half cycle to reduce momentary interruptions and voltage sags.

Today, the primary selective scheme is preferred mainly because of the cost associated with the extra transformer in a secondary selective scheme. The normally closed switch on the primary-side transfer switch opens after

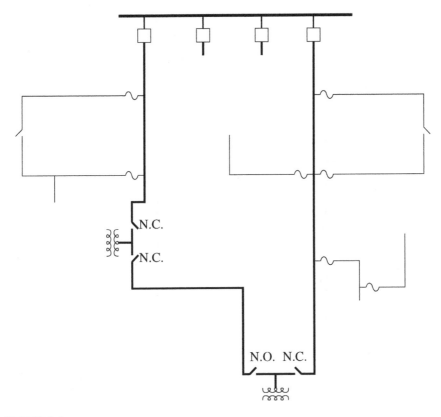

FIGURE 1.6
Primary loop distribution arrangement. (Copyright © 2000. Electric Power Research Institute. 1000419. *Engineering Guide for Integration of Distributed Generation and Storage Into Power Distribution Systems*. Reprinted with permission.)

sensing a loss of voltage. It normally has a time delay on the order of seconds — enough to ride through the distribution circuit's normal reclosing cycle. The opening of the switch is blocked if there is an overcurrent in the switch (the switch doesn't have fault interrupting capability). Transfer is also disabled if the alternate feed does not have proper voltage. The switch can return to normal through either an open or a closed transition; in a closed transition, both distribution circuits are temporarily paralleled.

1.2 Urban Networks

Some distribution circuits are not radial. The most common are the grid and spot secondary networks. In these systems, the secondary is networked together and has feeds from several primary distribution circuits. The spot

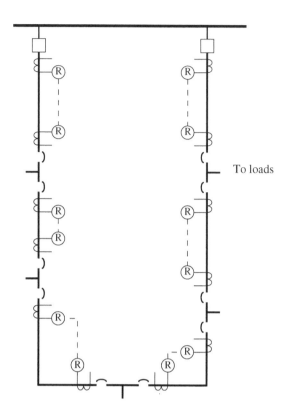

FIGURE 1.7
Example of a closed-loop distribution system.

network feeds one load such as a high-rise building. The grid network feeds several loads at different points in an area. Secondary networks are very reliable; if any of the primary distribution circuits fail, the others will carry the load without causing an outage for any customers.

The spot network generally is fed by three to five primary feeders (see Figure 1.9). The circuits are generally sized to be able to carry all of the load with the loss of either one or two of the primary circuits. Secondary networks have network protectors between the primary and the secondary network. A network protector is a low-voltage circuit breaker that will open when there is reverse power through it. When a fault occurs on a primary circuit, fault current backfeeds from the secondary network(s) to the fault. When this occurs, the network protectors will trip on reverse power. A spot network operates at 480Y/277 V or 208Y/120 V in the U.S.

Secondary grid networks are distribution systems that are used in most major cities. The secondary network is usually 208Y/120 V in the U.S. Five to ten primary distribution circuits (e.g., 12.47-kV circuits) feed the secondary network at multiple locations. Figure 1.10 shows a small part of a secondary network. As with a spot network, network protectors provide protection for faults on the primary circuits. Secondary grid networks can have peak loads

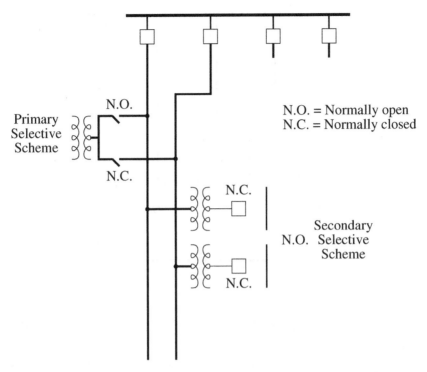

FIGURE 1.8
Primary and secondary selective schemes. (Copyright © 2000. Electric Power Research Institute. 1000419. *Engineering Guide for Integration of Distributed Generation and Storage Into Power Distribution Systems.* Reprinted with permission.)

of 5 to 50 MVA. Most utilities limit networks to about 50 MVA, but some networks are over 250 MVA. Loads are fed by tapping into the secondary networks at various points. Grid networks (also called street networks) can supply residential or commercial loads, either single or three phase. For single-phase loads, three-wire service is provided to give 120 V and 208 V (rather than the standard three-wire residential service, which supplies 120 V and 240 V).

Networks are normally fed by feeders originating from one substation bus. Having one source reduces circulating current and gives better load division and distribution among circuits. It also reduces the chance that network protectors stay open under light load (circulating current can trip the protectors). Given these difficulties, it is still possible to feed grid or spot networks from different substations or electrically separate buses.

The network protector is the key to automatic isolation and continued operation. The network protector is a three-phase low-voltage air circuit breaker with controls and relaying. The network protector is mounted on the network transformer or on a vault wall. Standard units are available with continuous ratings from 800 to 5000 A. Smaller units can interrupt 30 kA symmetrical, and larger units have interrupt ratings of 60 kA (IEEE Std.

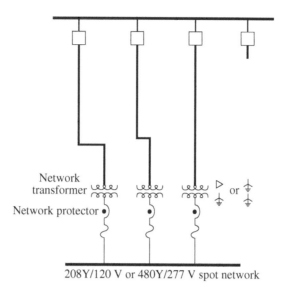

FIGURE 1.9
Spot network. (Copyright © 2000. Electric Power Research Institute. 1000419. *Engineering Guide for Integration of Distributed Generation and Storage Into Power Distribution Systems.* Reprinted with permission.)

C57.12.44–2000). A network protector senses and operates for reverse power flow (it does not have forward-looking protection). Protectors are available for either 480Y/277 V or 216Y/125 V.

The tripping current on network protectors can be changed, with low, nominal, and high settings, which are normally 0.05 to 0.1%, 0.15 to 0.20%, and 3 to 5% of the network protector rating. For example, a 2000-A network protector has a low setting of 1 A, a nominal setting of 4 A, and a high setting of 100 A (IEEE Std. C57.12.44–2000). Network protectors also have fuses that provide backup in case the network protector fails to operate, and as a secondary benefit, provide protection to the network protector and transformer against faults in the secondary network that are close.

The closing voltages are also adjustable: a 216Y/125-V protector has low, medium, and high closing voltages of 1 V, 1.5 V, and 2 V, respectively; a 480Y/277-V protector has low, medium, and high closing voltages of 2.2 V, 3.3 V, and 4.4 V, respectively.

1.3 Primary Voltage Levels

Most distribution voltages are between 4 and 35 kV. In this book, unless otherwise specified, voltages are given as line-to-line voltages; this follows normal industry practice, but it is sometimes a source of confusion. The four

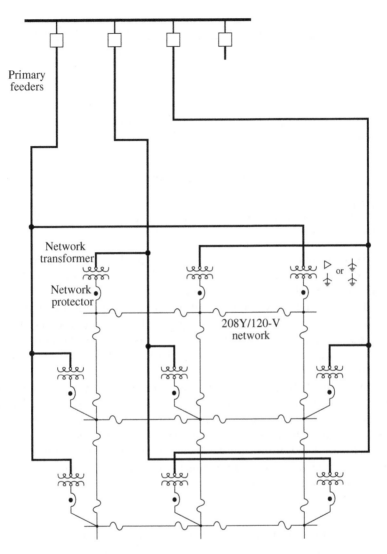

FIGURE 1.10
Portion of a grid network. (Copyright © 2000. Electric Power Research Institute. 1000419. *Engineering Guide for Integration of Distributed Generation and Storage Into Power Distribution Systems.* Reprinted with permission.)

major voltage classes are 5, 15, 25, and 35 kV. A voltage class is a term applied to a set of distribution voltages and the equipment common to them; it is not the actual system voltage. For example, a 15-kV insulator is suitable for application on any 15-kV class voltage, including 12.47 kV, 13.2 kV, and 13.8 kV. Cables, terminations, insulators, bushings, reclosers, and cutouts all have a voltage class rating. Only voltage-sensitive equipment like surge arresters, capacitors, and transformers have voltage ratings dependent on the actual system voltage.

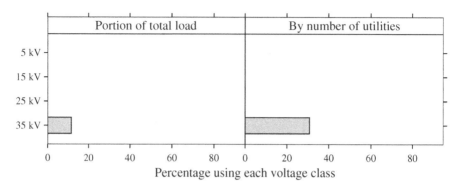

FIGURE 1.11
Usage of different distribution voltage classes (n = 107). (Data from [IEEE Working Group on Distribution Protection, 1995].)

Utilities most widely use the 15-kV voltages as shown by the survey results of North American utilities in Figure 1.11. The most common 15-kV voltage is 12.47 kV, which has a line-to-ground voltage of 7.2 kV.

The dividing line between distribution and subtransmission is often gray. Some lines act as both subtransmission and distribution circuits. A 34.5-kV circuit may feed a few 12.5-kV distribution substations, but it may also serve some load directly. Some utilities would refer to this as subtransmission, others as distribution.

The last half of the 20th century saw a move to higher voltage primary distribution systems. Higher-voltage distribution systems have advantages and disadvantages (see Table 1.3 for a summary). The great advantage of higher voltage systems is that they carry more power for a given current (Table 1.4 shows maximum power levels typically supplied by various distribution voltages). Less current means lower voltage drop, fewer losses, and more power-carrying capability. Higher voltage systems need fewer voltage

TABLE 1.3

Advantages and Disadvantages of Higher Voltage Distribution

Advantages	Disadvantages
Voltage drop — A higher-voltage circuit has less voltage drop for a given power flow.	*Reliability* — An important disadvantage of higher voltages: longer circuits mean more customer interruptions.
Capacity — A higher-voltage system can carry more power for a given ampacity.	*Crew safety and acceptance* — Crews do not like working on higher-voltage distribution systems.
Losses — For a given level of power flow, a higher-voltage system has fewer line losses.	*Equipment cost* — From transformers to cable to insulators, higher-voltage equipment costs more.
Reach — With less voltage drop and more capacity, higher voltage circuits can cover a much wider area.	
Fewer substations — Because of longer reach, higher-voltage distribution systems need fewer substations.	

TABLE 1.4

Power Supplied by Each Distribution
Voltage for a Current of 400 A

System Voltage (kV)	Total Power (MVA)
4.8	3.3
12.47	8.6
22.9	15.9
34.5	23.9

regulators and capacitors for voltage support. Utilities can use smaller conductors on a higher voltage system or carry more power on the same size conductor. Utilities can run much longer distribution circuits at a higher primary voltage, which means fewer distribution substations. Some fundamental relationships are:

- *Power* — For the same current, power changes linearly with voltage.

$$P_2 = \frac{V_1}{V_1} P_1$$

when $I_2 = I_1$

- *Current* — For the same power, increasing the voltage decreases current linearly.

$$I_2 = \frac{V_1}{V_2} I_1$$

when $P_2 = P_1$

- *Voltage drop* — For the same power delivered, the percentage voltage drop changes as the ratio of voltages squared. A 12.47-kV circuit has four times the percentage voltage drop as a 24.94-kV circuit carrying the same load.

$$V_{\%2} = \left(\frac{V_1}{V_2}\right)^2 V_{\%1}$$

when $P_2 = P_1$

- *Area coverage* — For the same load density, the area covered increases linearly with voltage: A 24.94-kV system can cover twice the area of a 12.47-kV system; a 34.5-kV system can cover 2.8 times the area of a 12.47-kV system.

$$A_2 = \frac{V_2}{V_1} A_1$$

where

V_1, V_2 = voltage on circuits 1 and 2

P_1, P_2 = power on circuits 1 and 2

I_1, I_2 = current on circuits 1 and 2

$V_{\%1}, V_{\%2}$ = voltage drop per unit length in percent on circuits 1 and 2

A_1, A_2 = area covered by circuits 1 and 2

The squaring effect on voltage drop is significant. It means that doubling the system voltage quadruples the load that can be supplied over the same distance (with equal percentage voltage drop); or, twice the load can be supplied over twice the distance; or, the same load can be supplied over four times the distance.

Resistive line losses are also lower on higher-voltage systems, especially in a voltage-limited circuit. Thermally limited systems have more equal losses, but even in this case higher voltage systems have fewer losses.

Line crews do not like higher voltage distribution systems as much. In addition to the widespread perception that they are not as safe, gloves are thicker, and procedures are generally more stringent. Some utilities will not glove 25- or 35-kV voltages and only use hotsticks.

The main disadvantage of higher-voltage systems is reduced reliability. Higher voltages mean longer lines and more exposure to lightning, wind, dig-ins, car crashes, and other fault causes. A 34.5-kV, 30-mi mainline is going to have many more interruptions than a 12.5-kV system with an 8-mi mainline. To maintain the same reliability as a lower-voltage distribution system, a higher-voltage primary must have more switches, more automation, more tree trimming, or other reliability improvements. Higher voltage systems also have more voltage sags and momentary interruptions. More exposure causes more momentary interruptions. Higher voltage systems have more voltage sags because faults further from the substation can pull down the station's voltage (on a higher voltage system the line impedance is lower relative to the source impedance).

Cost comparison between circuits is difficult (see Table 1.5 for one utility's cost comparison). Higher voltage equipment costs more — cables, insulators, transformers, arresters, cutouts, and so on. But higher voltage circuits can use smaller conductors. The main savings of higher-voltage distribution is fewer substations. Higher voltage systems also have lower annual costs from losses. As far as ongoing maintenance, higher voltage systems require less substation maintenance, but higher voltage systems should have more tree trimming and inspections to maintain reliability.

Conversion to a higher voltage is an option for providing additional capacity in an area. Conversion to higher voltages is most beneficial when substation

TABLE 1.5

Costs of 34.5 kV Relative to 12.5 kV

Item	Underground	Overhead
Subdivision without bulk feeders	1.25	1.13
Subdivision with bulk feeders	1.00	0.85
Bulk feeders	0.55	0.55
Commercial areas	1.05–1.25	1.05–1.25

Source: Jones, A.I., Smith, B.E., and Ward, D.J., "Considerations for Higher Voltage Distribution," *IEEE Transactions on Power Delivery*, vol. 7, no. 2, pp. 782–8, April 1992.

space is hard to find and load growth is high. If the existing subtransmission voltage is 34.5 kV, then using that voltage for distribution is attractive; additional capacity can be met by adding customers to existing 34.5-kV lines (a neutral may need to be added to the 34.5-kV subtransmission line).

Higher voltage systems are also more prone to ferroresonance. Radio interference is also more common at higher voltages.

Overall, the 15-kV class voltages provide a good balance between cost, reliability, safety, and reach. Although a 15-kV circuit does not naturally provide long reach, with voltage regulators and feeder capacitors it can be stretched to reach 20 mi or more. That said, higher voltages have advantages, especially for rural lines and for high-load areas, particularly where substation space is expensive.

Many utilities have multiple voltages (as shown by the survey data in Figure 1.11). Even one circuit may have multiple voltages. For example, a utility may install a 12.47-kV circuit in an area presently served by 4.16 kV. Some of the circuit may be converted to 12.47 kV, but much of it can be left as is and coupled through 12.47/4.16-kV step-down transformer banks.

1.4 Distribution Substations

Distribution substations come in many sizes and configurations. A small rural substation may have a nominal rating of 5 MVA while an urban station may be over 200 MVA. Figure 1.12 through Figure 1.14 show examples of small, medium, and large substations. As much as possible, many utilities have standardized substation layouts, transformer sizes, relaying systems, and automation and SCADA (supervisory control and data acquisition) facilities. Most distribution substation bus configurations are simple with limited redundancy.

Transformers smaller than 10 MVA are normally protected with fuses, but fuses are also used for transformers to 20 or 30 MVA. Fuses are inexpensive and simple; they don't need control power and take up little space. Fuses are not particularly sensitive, especially for evolving internal faults. Larger transformers normally have relay protection that operates a circuit switcher

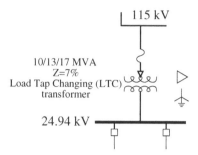

FIGURE 1.12
Example rural distribution substation.

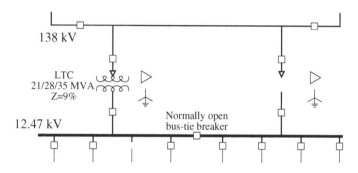

FIGURE 1.13
Example suburban distribution substation.

or a circuit breaker. Relays often include differential protection, sudden-pressure relays, and overcurrent relays. Both the differential protection and the sudden-pressure relays are sensitive enough to detect internal failures and clear the circuit to limit additional damage to the transformer. Occasionally, relays operate a high-side grounding switch instead of an interrupter. When the grounding switch engages, it creates a bolted fault that is cleared by an upstream device or devices.

The feeder interrupting devices are normally relayed circuit breakers, either free-standing units or metal-enclosed switchgear. Many utilities also use reclosers instead of breakers, especially at smaller substations.

Station transformers are normally protected by differential relays which trip if the current into the transformer is not very close to the current out of the transformer. Relaying may also include pressure sensors. The high-side protective device is often a circuit switcher but may also be fuses or a circuit breaker.

Two-bank stations are very common (Figure 1.13); these are the standard design for many utilities. Normally, utilities size the transformers so that if either transformer fails, the remaining unit can carry the entire substation's load. Utility practices vary on how much safety margin is built into this calculation, and load growth can eat into the redundancy.

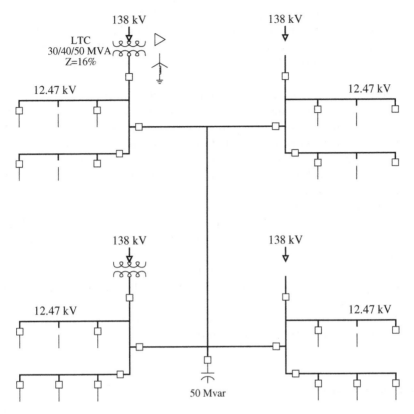

FIGURE 1.14
Example urban distribution substation.

Most utilities normally use a split bus: a bus tie between the two buses is normally left open in distribution substations. The advantages of a split bus are:

- *Lower fault current* — This is the main reason that bus ties are open. For a two-bank station with equal transformers, opening the bus tie cuts fault current in half.
- *Circulating current* — With a split bus, current cannot circulate through both transformers.
- *Bus regulation* — Bus voltage regulation is also simpler with a split bus. With the tie closed, control of paralleled tap changers is more difficult.

Having the bus tie closed has some advantages, and many utilities use closed ties under some circumstances. A closed bus tie is better for

- *Secondary networks* — When feeders from each bus supply either spot or grid secondary networks, closed bus ties help prevent circulating current through the secondary networks.

- *Unequal loading* — A closed bus tie helps balance the loading on the transformers. If the set of feeders on one bus has significantly different loading patterns (either seasonal or daily), then a closed bus tie helps even out the loading (and aging) of the two transformers.

Whether the bus tie is open or closed has little impact on reliability. In the uncommon event that one transformer fails, both designs allow the station to be reconfigured so that one transformer supplies both bus feeders. The closed-tie scenario is somewhat better in that an automated system can reconfigure the ties without total loss of voltage to customers (customers do see a very large voltage sag). In general, both designs perform about the same for voltage sags.

Urban substations are more likely to have more complicated bus arrangements. These could include ring buses or breaker-and-a-half schemes. Figure 1.14 shows an example of a large urban substation with feeders supplying secondary networks. If feeders are supplying secondary networks, it is not critical to maintain continuity to each feeder, but it is important to prevent loss of any one bus section or piece of equipment from shutting down the network (an *N*-1 design).

For more information on distribution substations, see (RUS 1724E-300, 2001; Westinghouse Electric Corporation, 1965).

1.5   Subtransmission Systems

Subtransmission systems are those circuits that supply distribution substations. Several different subtransmission systems can supply distribution substations. Common subtransmission voltages include 34.5, 69, 115, and 138 kV. Higher voltage subtransmission lines can carry more power with less losses over greater distances. Distribution circuits are occasionally supplied by high-voltage transmission lines such as 230 kV; such high voltages make for expensive high-side equipment in a substation. Subtransmission circuits are normally supplied by bulk transmission lines at subtransmission substations. For some utilities, one transmission system serves as both the subtransmission function (feeding distribution substations) and the transmission function (distributing power from bulk generators). There is much crossover in functionality and voltage. One utility may have a 23-kV subtransmission system supplying 4-kV distribution substations. Another utility right next door may have a 34.5-kV distribution system fed by a 138-kV subtransmission system. And within utilities, one can find a variety of different voltage combinations.

Of all of the subtransmission circuit arrangements, a radial configuration is the simplest and least expensive (see Figure 1.15). But radial circuits provide the most unreliable supply; a fault on the subtransmission circuit

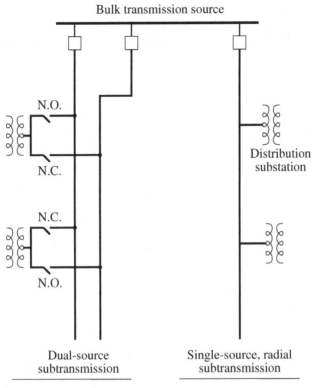

Bulk transmission source

N.O.

N.C.

N.C.

N.O.

Distribution substation

Dual-source subtransmission

More reliable: Faults on one of the radial subtransmission circuits should not cause interruptions to substations. Double-circuit faults can cause multiple station interruptions.

Single-source, radial subtransmission

Least reliable: Faults on the radial subtransmission circuit can cause interruptions to multiple substations.

FIGURE 1.15
Radial subtransmission systems.

can force an interruption of several distribution substations and service to many customers. A variety of redundant subtransmission circuits are available, including dual circuits and looped or meshed circuits (see Figure 1.16). The design (and evolution) of subtransmission configurations depends on how the circuit developed, where the load is needed now and in the future, what the distribution circuit voltages are, where bulk transmission is available, where rights-of-way are available, and, of course, economic factors.

Most subtransmission circuits are overhead. Many are built right along roads and streets just like distribution lines. Some — especially higher voltage subtransmission circuits — use a private right-of-way such as bulk transmission lines use. Some new subtransmission lines are put underground, as development of solid-insulation cables has made costs more reasonable.

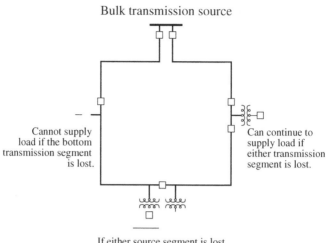

Bulk transmission source

Cannot supply
load if the bottom
transmission segment
is lost.

Can continue to
supply load if
either transmission
segment is lost.

If either source segment is lost,
one transformer can supply both
distribution buses.

FIGURE 1.16
Looped subtransmission system.

Lower voltage subtransmission lines (69, 34.5, and 23 kV) tend to be designed and operated as are distribution lines, with radial or simple loop arrangements, using wood-pole construction along roads, with reclosers and regulators, often without a shield wire, and with time-overcurrent protection. Higher voltage transmission lines (115, 138, and 230 kV) tend to be designed and operated like bulk transmission lines, with loop or mesh arrangements, tower configurations on a private right-of-way, a shield wire or wires for lightning protection, and directional or pilot-wire relaying from two ends. Generators may or may not interface at the subtransmission level (which can affect protection practices).

1.6   Differences between European and North American Systems

Distribution systems around the world have evolved into different forms. The two main designs are North American and European. This book deals mainly with North American distribution practices; for more information on European systems, see Lakervi and Holmes (1995). For both forms, hardware is much the same: conductors, cables, insulators, arresters, regulators, and transformers are very similar. Both systems are radial, and voltages and power carrying capabilities are similar. The main differences are in layouts, configurations, and applications.

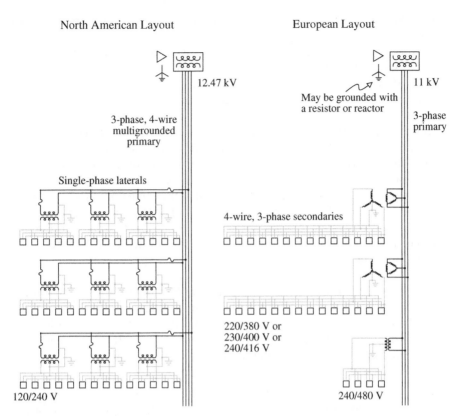

FIGURE 1.17
North American versus European distribution layouts.

Figure 1.17 compares the two systems. Relative to North American designs, European systems have larger transformers and more customers per transformer. Most European transformers are three-phase and on the order of 300 to 1000 kVA, much larger than typical North American 25- or 50-kVA single-phase units.

Secondary voltages have motivated many of the differences in distribution systems. North America has standardized on a 120/240-V secondary system; on these, voltage drop constrains how far utilities can run secondaries, typically no more than 250 ft. In European designs, higher secondary voltages allow secondaries to stretch to almost 1 mi. European secondaries are largely three-phase and most European countries have a standard secondary voltage of 220, 230, or 240 V, twice the North American standard. With twice the voltage, a circuit feeding the same load can reach four times the distance. And because three-phase secondaries can reach over twice the length of a single-phase secondary, overall, a European secondary can reach eight times the length of an American secondary for a given load and voltage drop. Although it is rare, some European utilities supply rural areas with single-phase taps made of two phases with single-phase transformers connected phase to phase.

In the European design, secondaries are used much like primary laterals in the North American design. In European designs, the primary is not tapped frequently, and primary-level fuses are not used as much. European utilities also do not use reclosing as religiously as North American utilities.

Some of the differences in designs center around the differences in loads and infrastructure. In Europe, the roads and buildings were already in place when the electrical system was developed, so the design had to "fit in." Secondary is often attached to buildings. In North America, many of the roads and electrical circuits were developed at the same time. Also, in Europe houses are packed together more and are smaller than houses in America.

Each type of system has its advantages. Some of the major differences between systems are the following (see also Carr and McCall, 1992; Meliopoulos et al., 1998; Nguyen et al., 2000):

- *Cost* — The European system is generally more expensive than the North American system, but there are so many variables that it is hard to compare them on a one-to-one basis. For the types of loads and layouts in Europe, the European system fits quite well. European primary equipment is generally more expensive, especially for areas that can be served by single-phase circuits.

- *Flexibility* — The North American system has a more flexible primary design, and the European system has a more flexible secondary design. For urban systems, the European system can take advantage of the flexible secondary; for example, transformers can be sited more conveniently. For rural systems and areas where load is spread out, the North American primary system is more flexible. The North American primary is slightly better suited for picking up new load and for circuit upgrades and extensions.

- *Safety* — The multigrounded neutral of the North American primary system provides many safety benefits; protection can more reliably clear faults, and the neutral acts as a physical barrier, as well as helping to prevent dangerous touch voltages during faults. The European system has the advantage that high-impedance faults are easier to detect.

- *Reliability* — Generally, North American designs result in fewer customer interruptions. Nguyen et al. (2000) simulated the performance of the two designs for a hypothetical area and found that the average frequency of interruptions was over 35% higher on the European system. Although European systems have less primary, almost all of it is on the main feeder backbone; loss of the main feeder results in an interruption for all customers on the circuit.

European systems need more switches and other gear to maintain the same level of reliability.

- *Power quality* — Generally, European systems have fewer voltage sags and momentary interruptions. On a European system, less primary exposure should translate into fewer momentary interruptions compared to a North American system that uses fuse saving. The three-wire European system helps protect against sags from line-to-ground faults. A squirrel across a bushing (from line to ground) causes a relatively high impedance fault path that does not sag the voltage much compared to a bolted fault on a well-grounded system. Even if a phase conductor faults to a low-impedance return path (such as a well-grounded secondary neutral), the delta – wye customer transformers provide better immunity to voltage sags, especially if the substation transformer is grounded through a resistor or reactor.

- *Aesthetics* — Having less primary, the European system has an aesthetic advantage: the secondary is easier to underground or to blend in. For underground systems, fewer transformer locations and longer secondary reach make siting easier.

- *Theft* — The flexibility of the European secondary system makes power much easier to steal. Developing countries especially have this problem. Secondaries are often strung along or on top of buildings; this easy access does not require great skill to attach into.

Outside of Europe and North America, both systems are used, and usage typically follows colonial patterns with European practices being more widely used. Some regions of the world have mixed distribution systems, using bits of North American and bits of European practices. The worst mixture is 120-V secondaries with European-style primaries; the low-voltage secondary has limited reach along with the more expensive European primary arrangement.

Higher secondary voltages have been explored (but not implemented to my knowledge) for North American systems to gain flexibility. Higher secondary voltages allow extensive use of secondary, which makes undergrounding easier and reduces costs. Westinghouse engineers contended that both 240/480-V three-wire single-phase and 265/460-V four-wire three-phase secondaries provide cost advantages over a similar 120/240-V three-wire secondary (Lawrence and Griscom, 1956; Lokay and Zimmerman, 1956). Higher secondary voltages do not force higher utilization voltages; a small transformer at each house converts 240 or 265 V to 120 V for lighting and standard outlet use (air conditioners and major appliances can be served directly without the extra transformation). More recently, Bergeron et al. (2000) outline a vision of a distribution system where primary-level distribution voltage is stepped down to an extensive 600-V, three-phase secondary system. At each house, an electronic transformer converts 600 V to 120/240 V.

1.7 Loads

Distribution systems obviously exist to supply electricity to end users, so loads and their characteristics are important. Utilities supply a broad range of loads, from rural areas with load densities of 10 kVA/mi^2 to urban areas with 300 MVA/mi^2. A utility may feed houses with a 10- to 20-kVA peak load on the same circuit as an industrial customer peaking at 5 MW. The electrical load on a feeder is the sum of all individual customer loads. And the electrical load of a customer is the sum of the load drawn by the customer's individual appliances. Customer loads have many common characteristics. Load levels vary through the day, peaking in the afternoon or early evening. Several definitions are used to quantify load characteristics at a given location on a circuit:

- *Demand* — The load average over a specified time period, often 15, 20, or 30 min. Demand can be used to characterize real power, reactive power, total power, or current. Peak demand over some period of time is the most common way utilities quantify a circuit's load. In substations, it is common to track the current demand.

- *Load factor* — The ratio of the average load over the peak load. Peak load is normally the maximum demand but may be the instantaneous peak. The load factor is between zero and one. A load factor close to 1.0 indicates that the load runs almost constantly. A low load factor indicates a more widely varying load. From the utility point of view, it is better to have high load-factor loads. Load factor is normally found from the total energy used (kilowatt-hours) as:

$$LF = \frac{kWh}{d_{kW} \times h}$$

where

LF = load factor

kWh = energy use in kilowatt-hours

d_{kW} = peak demand in kilowatts

h = number of hours during the time period

- *Coincident factor* — The ratio of the peak demand of a whole system to the sum of the individual peak demands within that system. The peak demand of the whole system is referred to as the peak *diversified* demand or as the peak *coincident* demand. The individual peak demands are the *noncoincident* demands. The coincident factor is less than or equal to one. Normally, the coincident factor is much less than one because each of the individual loads do not hit their peak at the same time (they are not coincident).

- *Diversity factor* — The ratio of the sum of the individual peak demands in a system to the peak demand of the whole system. The diversity factor is greater than or equal to one and is the reciprocal of the coincident factor.

- *Responsibility factor* — The ratio of a load's demand at the time of the system peak to its peak demand. A load with a responsibility factor of one peaks at the same time as the overall system. The responsibility factor can be applied to individual customers, customer classes, or circuit sections.

The loads of certain customer classes tend to vary in similar patterns. Commercial loads are highest from 8 a.m. to 6 p.m. Residential loads peak in the evening. Weather significantly changes loading levels. On hot summer days, air conditioning increases the demand and reduces the diversity among loads. At the transformer level, load factors of 0.4 to 0.6 are typical (Gangel and Propst, 1965).

Several groups have evaluated coincidence factors as a function of the number of customers. Nickel and Braunstein (1981) determined that one curve fell roughly in the middle of several curves evaluated. Used by Arkansas Power and Light, this curve fits the following:

$$F_{co} = \frac{1}{2}\left(1 + \frac{5}{2n + 3}\right)$$

where n is the number of customers (see Figure 1.18).

At the substation level, coincidence is also apparent. A transformer with four feeders, each peaking at 100 A, will peak at less than 400 A because of diversity between feeders. The coincident factor between four feeders is normally higher than coincident factors at the individual customer level. Expect coincident factors to be above 0.9. Each feeder is already highly diversified, so not much more is gained by grouping more customers together if the sets of customers are similar. If the customer mix on each feeder is different, then multiple feeders can have significant differences. If some feeders are mainly residential and others are commercial, the peak load of the feeders together can be significantly lower than the sum of the peaks. For distribution transformers, the peak responsibility factor ranges from 0.5 to 0.9 with 0.75 being typical (Nickel and Braunstein, 1981).

Different customer classes have different characteristics (see Figure 1.19 for an example). Residential loads peak more in the evening and have a

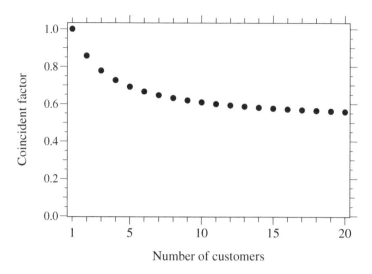

FIGURE 1.18
Coincident factor average curve for utilities.

relatively low load factor. Commercial loads tend to be more 8 a.m. to 6 p.m., and the industrial loads tend to run continuously and, as a class, they have a higher load factor.

1.8   The Past and the Future

Looking at Seelye's *Electrical Distribution Engineering* book (1930), we find more similarities to than differences from present-day distribution systems. The basic layout and operations of distribution infrastructure at the start of the 21st century are much the same as in the middle of the 20th century. Equipment has undergone steady improvements; transformers are more efficient; cables are much less expensive and easier to use; and protection equipment is better (see Figure 1.20 for some development milestones). Utilities operate more distribution circuits at higher voltages and use more underground circuits. But the concepts are much the same: ac, three-phase systems, radial circuits, fused laterals, overcurrent relays, etc. Advances in computer technology have opened up possibilities for more automation and more effective protection.

How will future distribution systems evolve? Given the fact that distribution systems of the year 2000 look much the same as distribution systems in 1950, a good guess is that the distribution system of 2050 (or at least 2025) will look much like today's systems. More and more of the electrical infrastructure will be placed underground. Designs and equipment will continue

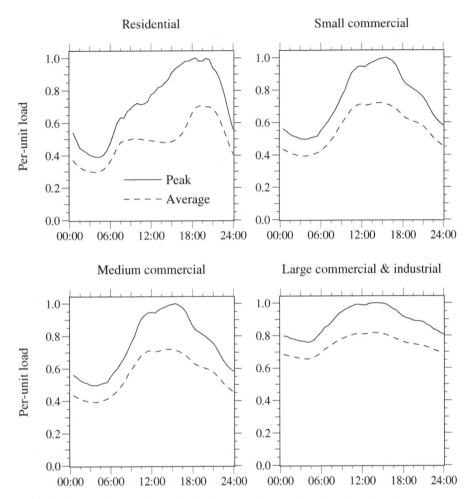

FIGURE 1.19
Daily load profiles for Pacific Gas and Electric (2002 data).

to be standardized. Gradually, the distribution system will evolve to take advantage of computer and communication gains: more automation, more communication between equipment, and smarter switches and controllers. EPRI outlined a vision of a future distribution system that was no longer radial, a distribution system that evolves to support widespread distributed generation and storage along with the ability to charge electric vehicles (EPRI TR-111683, 1998). Such a system needs directional relaying for reclosers, communication between devices, regulators with advanced controls, and information from and possibly control of distributed generators.

Advances in power electronics make more radical changes such as conversion to dc possible. Advances in power electronics allow flexible conversion between different frequencies, phasings, and voltages while still

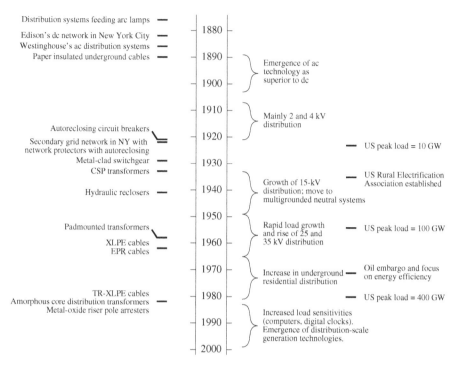

FIGURE 1.20
Electric power distribution development timeline.

producing ac voltage to the end user at the proper voltage. While possible, radical changes are unlikely, given the advantages to evolving an existing system rather than replacing it. Whatever the approach, the future has challenges; utilities will be expected to deliver more reliable power with minimal pollution while keeping the distribution system hidden from view and causing the least disruption possible. And of course, costs are expected to stay the same or go down.

References

Bergeron, R., Slimani, K., Lamarche, L., and Cantin, B., "New Architecture of the Distribution System Using Electronic Transformer," ESMO-2000, Panel on Distribution Transformer, Breakers, Switches and Arresters, 2000.

Carr, J. and McCall, L.V., "Divergent Evolution and Resulting Characteristics Among the World's Distribution Systems," *IEEE Transactions on Power Delivery*, vol. 7, no. 3, pp. 1601–9, July 1992.

CEA, *CEA Distribution Planner's Manual*, Canadian Electrical Association, 1982.

EPRI 1000419, *Engineering Guide for Integration of Distributed Generation and Storage Into Power Distribution Systems*, Electric Power Research Institute, Palo Alto, CA, 2000.

EPRI TR-109178, *Distribution Cost Structure — Methodology and Generic Data*, Electric Power Research Institute, Palo Alto, CA, 1998.

EPRI TR-111683, *Distribution Systems Redesign*, Electric Power Research Institute, Palo Alto, CA, 1998.

Gangel, M.W. and Propst, R.F., "Distribution Transformer Load Characteristics," *IEEE Transactions on Power Apparatus and Systems*, vol. 84, pp. 671–84, August 1965.

IEEE Std. C57.12.44–2000, IEEE Standard Requirements for Secondary Network Protectors.

IEEE Tutorial Course, *Power Distribution Planning*, 1992. Course text 92 EHO 361–6-PWR.

IEEE Working Group on Distribution Protection, "Distribution Line Protection Practices Industry Survey Results," *IEEE Transactions on Power Delivery*, vol. 10, no. 1, pp. 176–86, January 1995.

Jones, A.I., Smith, B.E., and Ward, D.J., "Considerations for Higher Voltage Distribution," *IEEE Transactions on Power Delivery*, vol. 7, no. 2, pp. 782–8, April 1992.

Lakervi, E. and Holmes, E.J., *Electricity Distribution Network Design*, IEE Power Engineering Series 21, Peter Peregrinius, 1995.

Lawrence, R.F. and Griscom, S.B., "Residential Distribution — An Analysis of Systems to Serve Expanding Loads," *AIEE Transactions, Part III*, vol. 75, pp. 533–42, 1956.

Lokay, H.E. and Zimmerman, R.A., "Economic Comparison of Secondary Voltages: Single and Three Phase Distribution for Residential Areas," *AIEE Transactions, Part III*, vol. 75, pp. 542–52, 1956.

Meliopoulos, A.P. S., Kennedy, J., Nucci, C.A., Borghetti, A., and Contaxis, G., "Power Distribution Practices in USA and Europe: Impact on Power Quality," 8th International Conference on Harmonics and Quality of Power, 1998.

Nguyen, H.V., Burke, J.J., and Benchluch, S., "Rural Distribution System Design Comparison," IEEE Power Engineering Society Winter Meeting, 2000.

Nickel, D.L. and Braunstein, H.R., "Distribution. Transformer Loss Evaluation. II. Load Characteristics and System Cost Parameters," *IEEE Transactions on Power Apparatus and Systems*, vol. PAS-100, no. 2, pp. 798–811, February 1981.

Pagel, B., "Energizing International Drive," *Transmission & Distribution World*, April 2000.

RUS 1724E-300, *Design Guide for Rural Substations*, United States Department of Agriculture, Rural Utilities Service, 2001.

Seelye, H.P., *Electrical Distribution Engineering*, McGraw-Hill New York, 1930.

Westinghouse Electric Corporation, *Distribution Systems*, vol. 3, 1965.

Willis, H.L., *Power Distribution Planning Reference Book*, Marcel Dekker, New York, 1997.

No matter how long you've been a Power Lineman, you still notice it when people refer to your poles as "telephone poles."

Powerlineman law #46, By CD Thayer and other Power Linemen,
http://www.cdthayer.com/lineman.htm

2

Overhead Lines

Along streets, alleys, through woods, and in backyards, many of the distribution lines that feed customers are overhead structures. Because overhead lines are exposed to trees and animals, to wind and lightning, and to cars and kites, they are a critical component in the reliability of distribution circuits. This chapter discusses many of the key electrical considerations of overhead lines: conductor characteristics, impedances, ampacity, and other issues.

2.1 Typical Constructions

Overhead constructions come in a variety of configurations (see Figure 2.1). Normally one primary circuit is used per pole, but utilities sometimes run more than one circuit per structure. For a three-phase circuit, the most common structure is a horizontal layout with an 8- or 10-ft wood crossarm on a pole (see Figure 2.2). Armless constructions are also widely found where fiberglass insulator standoffs or post insulators are used in a tighter configuration. Utilities normally use 30- to 45-ft poles, set 6 to 8 ft deep. Vertical construction is also occasionally used. Span lengths vary from 100 to 150 ft in suburban areas to as much as 300 or 400 ft in rural areas.

Distribution circuits normally have an underbuilt neutral — the neutral acts as a safety ground for equipment and provides a return path for unbalanced loads and for line-to-ground faults. The neutral is 3 to 5 ft below the phase conductors. Utilities in very high lightning areas may run the neutral wire above the phase conductors to act as a shield wire. Some utilities also run the neutral on the crossarm. Secondary circuits are often run under the primary. The primary and the secondary may share the neutral, or they may each have their own neutral. Many electric utilities share their space with other utilities; telephone or cable television cables may run under the electric secondary.

(a)

FIGURE 2.1
Example overhead distribution structures. (a) Three-phase 34.5-kV armless construction with covered wire.

(b)

FIGURE 2.1
Continued. (b) Single-phase circuit, 7.2 kV line-to-ground.

(c)

FIGURE 2.1
Continued. (c) Single-phase, 4.8-kV circuit.

(d)

FIGURE 2.1
Continued. (d) 13.2-kV spacer cable.

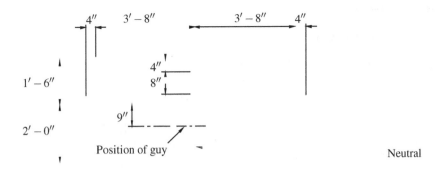

FIGURE 2.2
Example crossarm construction. (From [RUS 1728F-803, 1998].)

Wood is the main pole material, although steel, concrete, and fiberglass are also used. Treated wood lasts a long time, is easy to climb and attach equipment to, and also augments the insulation between the energized conductors and ground. Conductors are primarily aluminum. Insulators are pin type, post type, or suspension, either porcelain or polymer.

The National Electrical Safety Code (IEEE C2-2000) governs many of the safety issues that play important roles in overhead design issues. Poles must have space for crews to climb them and work safely in the air. All equipment must have sufficient strength to stand up to "normal" operations. Conductors must carry their weight, the weight of any accumulated ice, plus withstand the wind pressure exerted on the wire. We are not going to discuss mechanical and structural issues in this book. For more information, see the *Lineman's and Cableman's Handbook* (Kurtz et al., 1997), the *Mechanical Design Manual for Overhead Distribution Lines* (RUS 160-2, 1982), the *NESC* (IEEE C2-2000), and the *NESC Handbook* (Clapp, 1997).

Overhead construction can cost $10,000/mi to $250,000/mi, depending on the circumstances. Some of the major variables are labor costs, how developed the land is, natural objects (including rocks in the ground and trees in the way), whether the circuit is single or three phase, and how big the conductors are. Suburban three-phase mains are typically about $60,000 to $150,000/mi; single-phase laterals are often in the $40,000 to $75,000/mi range. Construction is normally less expensive in rural areas; in urban areas, crews must deal with traffic and set poles in concrete. As Willis (1997) notes, upgrading a circuit normally costs more than building a new line. Typically this work is done live: the old conductor has to be moved to standoff brackets while the new conductor is strung, and the poles may have to be reinforced to handle heavier conductors.

2.2   Conductor Data

A *wire* is metal drawn or rolled to long lengths, normally understood to be a solid wire. Wires may or may not be insulated. A *conductor* is one or more wires suitable for carrying electric current. Often the term *wire* is used to mean conductor. Table 2.1 shows some characteristics of common conductor metals.

Most conductors are either aluminum or copper. Utilities use aluminum for almost all new overhead installations. Aluminum is lighter and less expensive for a given current-carrying capability. Copper was installed more in the past, so significant lengths of copper are still in service on overhead circuits.

Aluminum for power conductors is alloy 1350, which is 99.5% pure and has a minimum conductivity of 61.0% IACS [for more complete characteristics, see the *Aluminum Electrical Conductor Handbook* (Aluminum Association, 1989)]. Pure aluminum melts at 660°C. Aluminum starts to anneal

TABLE 2.1

Nominal or Minimum Properties of Conductor Wire Materials

Property	International Annealed Copper Standard	Commercial Hard-Drawn Copper Wire	Standard 1350-H19 Aluminum Wire	Standard 6201-T81 Aluminum Wire	Galvanized Steel Core Wire	Aluminum Clad Steel
Conductivity,% IACS at 20°C	100.0	97.0	61.2	52.5	8.0	20.3
Resistivity at 20°C, $\Omega \cdot in.^2/$ 1000 ft	0.008145	0.008397	0.013310	0.015515	0.101819	0.04007
Ratio of weight for equal dc resistance and length	1.00	1.03	0.50	0.58	9.1	3.65
Temp. coefficient of resistance, per °C at 20°C	0.00393	0.00381	0.00404	0.00347	0.00327	0.00360
Density at 20°C, lb/in.^{3}	0.3212	0.3212	0.0977	0.0972	0.2811	0.2381
Coefficient of linear expansion, 10^{-6} per °C	16.9	16.9	23.0	23.0	11.5	13.0
Modulus of elasticity, 10^6 psi	17	17	10	10	29	23.5
Specific heat at 20°C, cal/gm-°C	0.0921	0.0921	0.214	0.214	0.107	0.112
Tensile strength, 10^3 psi	62.0	62.0	24.0	46.0	185	175
Minimum elongation,%	1.1	1.1	1.5	3.0	3.5	1.5

Source: Southwire Company, *Overhead Conductor Manual,* 1994.

(soften and lose strength) above 100°C. It has good corrosion resistance; when exposed to the atmosphere, aluminum oxidizes, and this thin, invisible film of aluminum oxide protects against most chemicals, weathering conditions, and even acids. Aluminum can corrode quickly through electrical contact with copper or steel. This galvanic corrosion (dissimilar metals corrosion) accelerates in the presence of salts.

Several variations of aluminum conductors are available:

- *AAC — all-aluminum conductor* — Aluminum grade 1350-H19 AAC has the highest conductivity-to-weight ratio of all overhead conductors. See Table 2.2 for characteristics.

- *ACSR — aluminum conductor, steel reinforced* — Because of its high mechanical strength-to-weight ratio, ACSR has equivalent or higher ampacity for the same size conductor (the kcmil size designation is determined by the cross-sectional area of the aluminum; the steel is neglected). The steel adds extra weight, normally 11 to 18% of the weight of the conductor. Several different strandings are available to provide different strength levels. Common distribution sizes of ACSR have twice the breaking strength of AAC. High strength means the conductor can withstand higher ice and wind loads. Also, trees are less likely to break this conductor. See Table 2.3 for characteristics.

- *AAAC — all-aluminum alloy conductor* — This alloy of aluminum, the 6201-T81 alloy, has high strength and equivalent ampacities of AAC or ACSR. AAAC finds good use in coastal areas where use of ACSR is prohibited because of excessive corrosion.

- *ACAR — aluminum conductor, alloy reinforced* — Strands of aluminum 6201-T81 alloy are used along with standard 1350 aluminum. The alloy strands increase the strength of the conductor. The strands of both are the same diameter, so they can be arranged in a variety of configurations.

For most urban and suburban applications, AAC has sufficient strength and has good thermal characteristics for a given weight. In rural areas, utilities can use smaller conductors and longer pole spans, so ACSR or another of the higher-strength conductors is more appropriate.

Copper has very low resistivity and is widely used as a power conductor, although use as an overhead conductor has become rare because copper is heavier and more expensive than aluminum. It has significantly lower resistance than aluminum by volume — a copper conductor has equivalent ampacity (resistance) of an aluminum conductor that is two AWG sizes larger. Copper has very good resistance to corrosion. It melts at 1083°C, starts to anneal at about 100°C, and anneals most rapidly between 200 and 325°C (this range depends on the presence of impurities and amount of hardening). When copper anneals, it softens and loses tensile strength.

TABLE 2.2

Characteristics of All-Aluminum Conductor (AAC)

AWG	kcmil	Strands	Diameter, in.	GMR, ft	Resistance, Ω/1000 ft					Breaking. Strength, lb	Weight, lb/1000 ft
					dc	60-Hz ac					
					20°C	25°C	50°C	75°C			
6	26.24	7	0.184	0.0056	0.6593	0.6725	0.7392	0.8059		563	24.6
4	41.74	7	0.232	0.0070	0.4144	0.4227	0.4645	0.5064		881	39.1
2	66.36	7	0.292	0.0088	0.2602	0.2655	0.2929	0.3182		1350	62.2
1	83.69	7	0.328	0.0099	0.2066	0.2110	0.2318	0.2527		1640	78.4
1/0	105.6	7	0.368	0.0111	0.1638	0.1671	0.1837	0.2002		1990	98.9
2/0	133.1	7	0.414	0.0125	0.1299	0.1326	0.1456	0.1587		2510	124.8
3/0	167.8	7	0.464	0.0140	0.1031	0.1053	0.1157	0.1259		3040	157.2
4/0	211.6	7	0.522	0.0158	0.0817	0.0835	0.0917	0.1000		3830	198.4
	250	7	0.567	0.0171	0.0691	0.0706	0.0777	0.0847		4520	234.4
	250	19	0.574	0.0181	0.0693	0.0706	0.0777	0.0847		4660	234.3
	266.8	7	0.586	0.0177	0.0647	0.0663	0.0727	0.0794		4830	250.2
	266.8	19	0.593	0.0187	0.0648	0.0663	0.0727	0.0794		4970	250.1
	300	19	0.629	0.0198	0.0575	0.0589	0.0648	0.0705		5480	281.4
	336.4	19	0.666	0.0210	0.0513	0.0527	0.0578	0.0629		6150	315.5
	350	19	0.679	0.0214	0.0494	0.0506	0.0557	0.0606		6390	327.9
	397.5	19	0.724	0.0228	0.0435	0.0445	0.0489	0.0534		7110	372.9
	450	19	0.769	0.0243	0.0384	0.0394	0.0434	0.0472		7890	421.8
	477	19	0.792	0.0250	0.0363	0.0373	0.0409	0.0445		8360	446.8
	477	37	0.795	0.0254	0.0363	0.0373	0.0409	0.0445		8690	446.8
	500	19	0.811	0.0256	0.0346	0.0356	0.0390	0.0426		8760	468.5
	500	37	0.813	0.0260	0.0346	0.0356	0.0390	0.0426		9110	468.3
	556.5	19	0.856	0.0270	0.0311	0.0320	0.0352	0.0383		9750	521.4
	556.5	37	0.858	0.0275	0.0311	0.0320	0.0352	0.0383		9940	521.3
	600	37	0.891	0.0285	0.0288	0.0297	0.0326	0.0356		10700	562.0
	636	37	0.918	0.0294	0.0272	0.0282	0.0309	0.0335		11400	596.0
	650	37	0.928	0.0297	0.0266	0.0275	0.0301	0.0324		11600	609.8
	700	37	0.963	0.0308	0.0247	0.0256	0.0280	0.0305		12500	655.7
	700	61	0.964	0.0310	0.0247	0.0256	0.0280	0.0305		12900	655.8
	715.5	37	0.974	0.0312	0.0242	0.0250	0.0275	0.0299		12800	671.0
	715.5	61	0.975	0.0314	0.0242	0.0252	0.0275	0.0299		13100	671.0
	750	37	0.997	0.0319	0.0230	0.0251	0.0263	0.0286		13100	703.2
	750	61	0.998	0.0321	0.0230	0.0251	0.0263	0.0286		13500	703.2
	795	37	1.026	0.0328	0.0217	0.0227	0.0248	0.0269		13900	745.3
	795	61	1.028	0.0331	0.0217	0.0227	0.0248	0.0269		14300	745.7
	874.5	37	1.077	0.0344	0.0198	0.0206	0.0227	0.0246		15000	820.3
	874.5	61	1.078	0.0347	0.0198	0.0206	0.0227	0.0246		15800	820.6
	900	37	1.092	0.0349	0.0192	0.0201	0.0220	0.0239		15400	844.0
	900	61	1.094	0.0352	0.0192	0.0201	0.0220	0.0239		15900	844.0
	954	37	1.124	0.0360	0.0181	0.0191	0.0208	0.0227		16400	894.5
	954	61	1.126	0.0362	0.0181	0.0191	0.0208	0.0225		16900	894.8
	1000	37	1.151	0.0368	0.0173	0.0182	0.0199	0.0216		17200	937.3
	1000	61	1.152	0.0371	0.0173	0.0182	0.0199	0.0216		17700	936.8

Source: Aluminum Association, *Aluminum Electrical Conductor Handbook*, 1989; Southwire Company, *Overhead Conductor Manual*, 1994.

Different sizes of conductors are specified with gage numbers or area in circular mils. Smaller wires are normally referred to using the American wire gage (AWG) system. The gage is a numbering scheme that progresses geo-metrically. A number 36 solid wire has a defined diameter of 0.005 in. (0.0127

TABLE 2.3

Characteristics of Aluminum Conductor, Steel Reinforced (ACSR)

| | | | | | Resistance, Ω/1000 ft | | | | Breaking. | |
| | | | | | dc | 60-Hz ac | | | Strength, | Weight, |
AWG	kcmil	Strands	Diameter, in.	GMR, ft	20°C	25°C	50°C	75°C	lb	lb/1000 ft
6	26.24	6/1	0.198	0.0024	0.6419	0.6553	0.7500	0.8159	1190	36.0
4	41.74	6/1	0.250	0.0033	0.4032	0.4119	0.4794	0.5218	1860	57.4
4	41.74	7/1	0.257	0.0045	0.3989	0.4072	0.4633	0.5165	2360	67.0
2	66.36	6/1	0.316	0.0046	0.2534	0.2591	0.3080	0.3360	2850	91.2
2	66.36	7/1	0.325	0.0060	0.2506	0.2563	0.2966	0.3297	3640	102
1	83.69	6/1	0.355	0.0056	0.2011	0.2059	0.2474	0.2703	3550	115
1/0	105.6	6/1	0.398	0.0071	0.1593	0.1633	0.1972	0.2161	4380	145
2/0	133.1	6/1	0.447	0.0077	0.1265	0.1301	0.1616	0.1760	5300	183
3/0	167.8	6/1	0.502	0.0090	0.1003	0.1034	0.1208	0.1445	6620	230
4/0	211.6	6/1	0.563	0.0105	0.0795	0.0822	0.1066	0.1157	8350	291
	266.8	18/1	0.609	0.0197	0.0644	0.0657	0.0723	0.0788	6880	289
	266.8	26/7	0.642	0.0217	0.0637	0.0652	0.0714	0.0778	11300	366
	336.4	18/1	0.684	0.0221	0.0510	0.0523	0.0574	0.0625	8700	365
	336.4	26/7	0.721	0.0244	0.0506	0.0517	0.0568	0.0619	14100	462
	336.4	30/7	0.741	0.0255	0.0502	0.0513	0.0563	0.0614	17300	526
	397.5	18/1	0.743	0.0240	0.0432	0.0443	0.0487	0.0528	9900	431
	397.5	26/7	0.783	0.0265	0.0428	0.0438	0.0481	0.0525	16300	546
	477	18/1	0.814	0.0263	0.0360	0.0369	0.0405	0.0441	11800	517
	477	24/7	0.846	0.0283	0.0358	0.0367	0.0403	0.0439	17200	614
	477	26/7	0.858	0.0290	0.0357	0.0366	0.0402	0.0438	19500	655
	477	30/7	0.883	0.0304	0.0354	0.0362	0.0389	0.0434	23800	746
	556.5	18/1	0.879	0.0284	0.0309	0.0318	0.0348	0.0379	13700	603
	556.5	24/7	0.914	0.0306	0.0307	0.0314	0.0347	0.0377	19800	716
	556.5	26/7	0.927	0.0313	0.0305	0.0314	0.0345	0.0375	22600	765
	636	24/7	0.977	0.0327	0.0268	0.0277	0.0300	0.0330	22600	818
	636	26/7	0.990	0.0335	0.0267	0.0275	0.0301	0.0328	25200	873
	795	45/7	1.063	0.0352	0.0216	0.0225	0.0246	0.0267	22100	895
	795	26/7	1.108	0.0375	0.0214	0.0222	0.0242	0.0263	31500	1093
	954	45/7	1.165	0.0385	0.0180	0.0188	0.0206	0.0223	25900	1075
	954	54/7	1.196	0.0404	0.0179	0.0186	0.0205	0.0222	33800	1228
	1033.5	45/7	1.213	0.0401	0.0167	0.0175	0.0191	0.0208	27700	1163

Sources: Aluminum Association, *Aluminum Electrical Conductor Handbook*, 1989; Southwire Company, *Overhead Conductor Manual*, 1994.

cm), and the largest size, a number 0000 (referred to as 4/0 and pronounced "four-ought") solid wire has a 0.46-in. (1.17-cm) diameter. The larger gage sizes in sequence of increasing conductor size are: 4, 3, 2, 1, 0 (1/0), 00 (2/0), 000 (3/0), 0000 (4/0). Going to the next bigger size (smaller gage number) increases the diameter by 1.1229. Some other useful rules are:

- An increase of three gage sizes doubles the area and weight and halves the dc resistance.
- An increase of six gage sizes doubles the diameter.

Larger conductors are specified in circular mils of cross-sectional area. One circular mil is the area of a circle with a diameter of one mil (one mil is one-thousandth of an inch). Conductor sizes are often given in kcmil, thousands of circular mils. In the past, the abbreviation MCM was used, which also means thousands of circular mils (M is thousands, not mega, in this case). By definition, a solid 1000-kcmil wire has a diameter of 1 in. The diameter of a solid wire in mils is related to the area in circular mils by $d = \sqrt{A}$.

Outside of America, most conductors are specified in mm². Some useful conversion relationships are:

$$1 \text{ kcmil} = 1000 \text{ cmil} = 785.4 \times 10^{-6} \text{ in}^2 = 0.5067 \text{ mm}^2$$

Stranded conductors increase flexibility. A two-layer arrangement has seven wires; a three-layer arrangement has 19 wires, and a four-layer arrangement has 37 wires. The cross-sectional area of a stranded conductor is the cross-sectional area of the metal, so a stranded conductor has a larger diameter than a solid conductor of the same area.

The area of an ACSR conductor is defined by the area of the aluminum in the conductor.

Utilities with heavy tree cover often use covered conductors — conductors with a thin insulation covering. The covering is not rated for full conductor line-to-ground voltage, but it is thick enough to reduce the chance of flash-over when a tree branch falls between conductors. Covered conductor is also called *tree wire* or weatherproof wire. Tree wire also helps with animal faults and allows utilities to use armless or candlestick designs or other tight configurations. Tree wire is available with a variety of covering types. The insulation materials polyethylene, XLPE, and EPR are common. Insulation thicknesses typically range from 30 to 150 mils (1 mil = 0.001 in. = 0.00254 cm); see Table 2.4 for typical thicknesses. From a design and operating viewpoint, covered conductors must be treated as bare conductors according to the National Electrical Safety Code (NESC) (IEEE C2-2000), with the only difference that tighter conductor spacings are allowed. It is also used in Australia to reduce the threat of bush fires (Barber, 1999).

While covered wire helps with trees, it has some drawbacks compared with bare conductors. Covered wire is much more susceptible to burndowns caused by fault arcs. Covered-wire systems increase the installed cost some-what. Covered conductors are heavier and have a larger diameter, so the ice and wind loading is higher than a comparable bare conductor. The covering may be susceptible to degradation due to ultraviolet radiation, tracking, and mechanical effects that cause cracking. Covered conductors are more sus-ceptible to corrosion, primarily from water. If water penetrates the covering, it settles at the low points and causes corrosion (it cannot evaporate). On bare conductors, corrosion is rare; rain washes bare conductors periodically, and evaporation takes care of moisture. The Australian experience has been that complete corrosion can occur with covered wires in 15 to 20 years of operation (Barber, 1999). Water enters the conductor at pinholes caused by lightning strikes, at cover damage caused by abrasion or erosion, and at holes pierced by connectors. Temperature changes then cause water to be

TABLE 2.4

Typical Covering Thicknesses of Covered All-Aluminum
Conductor

Size AWG or kcmil	Strands	Cover Thickness, mil	Diameter, in.	
			Bare	Covered
6	7	30	0.184	0.239
4	7	30	0.232	0.285
2	7	45	0.292	0.373
1	7	45	0.328	0.410
1/0	7	60	0.368	0.480
2/0	7	60	0.414	0.524
3/0	7	60	0.464	0.574
4/0	7	60	0.522	0.629
266.8	19	60	0.593	0.695
336.4	19	60	0.666	0.766
397.5	19	80	0.724	0.857
477	37	80	0.795	0.926
556.5	37	80	0.858	0.988
636	61	95	0.918	1.082
795	61	95	1.028	1.187

pumped into the conductor. Because of corrosion concerns, water-blocked
conductors are better.

Spacer cables and aerial cables are also alternatives that perform well in
treed areas. Spacer cables are a bundled configuration using a messenger
wire holding up three phase wires that use covered wire. Because the spacer
cable has significantly smaller spacings than normal overhead constructions,
its reactive impedance is smaller.

Guy wires, messenger wires, and other wires that require mechanical
strength but not current-carrying capability are often made of steel. Steel has
high strength (see Table 2.5). Steel corrodes quickly, so most applications use
galvanized steel to slow down the corrosion. Because steel is a magnetic
material, steel conductors also suffer hysteresis losses. Steel conductors have
much higher resistances than copper or aluminum. For some applications
requiring strength and conductivity, steel wires coated with copper or aluminum are available. A copperweld conductor has copper-coated steel
strands, and an alumoweld conductor aluminum-coated steel strands. Both
have better corrosion protection than galvanized steel.

2.3 Line Impedances

Overhead lines have resistance and reactance that impedes the flow of current. These impedance values are necessary for voltage drop, power flow,
short circuit, and line-loss calculations.

TABLE 2.5

Characteristics of Steel Conductors

| | | | | | | Resistance, Ω/1000 ft | | | | |
| | | | | | | | 60-Hz ac at the given current level | | | |
Size	Diameter, in.	Conductor Area, in.^{2}	Weight, lb/1000 ft	Strength, lb	dc 25°C	10A	40A	70A	100A
High-Strength Steel — Class A Galvanizing									
5/8	0.621	0.2356	813	29,600	0.41	0.42	0.43	0.46	0.49
1/2	0.495	0.1497	517	18,800	0.65	0.66	0.68	0.73	0.77
7/16	0.435	0.1156	399	14,500	0.84	0.85	0.88	0.94	1.00
3/8	0.360	0.0792	273	10,800	1.23	1.25	1.28	1.38	1.46
Utilities Grade Steel									
7/16	0.435	0.1156	399	18,000	0.87	0.88	0.90	0.95	1.02
3/8	0.380	0.0882	273	11,500	1.23	1.25	1.28	1.38	1.46
Extra-High-Strength Steel — Class A Galvanizing									
5/8	0.621	0.2356	813	42,400	0.43	0.43	0.44	0.47	0.50
1/2	0.495	0.1497	517	26,900	0.67	0.68	0.69	0.73	0.78
7/16	0.435	0.1156	399	20,800	0.87	0.88	0.90	0.95	1.02
3/8	0.360	0.0792	273	15,400	1.28	1.29	1.31	1.39	1.48
Extra-High-Strength Steel — Class C Galvanizing									
7/16	0.435	0.1156	399	20,800		0.70	0.70	0.71	0.71
3/8	0.360	0.0792	273	15,400		1.03	1.03	1.03	1.04
5/16	0.312	0.0595	205	11,200		1.20	1.30	1.30	1.30

Source: EPRI, *Transmission Line Reference Book: 345 kV and Above*, 2nd ed., Electric Power Research Institute, Palo Alto, CA, 1982.

The dc resistance is inversely proportional to the area of a conductor; doubling the area halves the resistance. Several units are used to describe a conductor's resistance. Conductivity is often given as %IACS, the percent conductivity relative to the International Annealed Copper Standard, which has the following volume resistivities:

$$0.08145 \ \Omega\text{-in.}^2/1000 \text{ ft} = 17.241 \ \Omega\text{-mm}^2/\text{km} = 10.37 \ \Omega\text{-cmil}/\text{ft}$$

And with a defined density of 8.89 g/cm^3 at 20°C, the copper standard has the following weight resistivities:

$$875.2 \ \Omega\text{-lb}/\text{mi}^2 = 0.15328 \ \Omega\text{-g}/\text{m}^2$$

Hard-drawn copper has 97.3%IACS. Aluminum varies, depending on type; alloy 1350-H19 has 61.2% conductivity.

Temperature and frequency — these change the resistance of a conductor. A hotter conductor provides more resistance to the flow of current. A higher frequency increases the internal magnetic fields. Current has a difficult time flowing in the center of a conductor at high frequency, as it is being opposed by the magnetic field generated by current flowing on all sides of it. Current flows more easily near the edges. This *skin effect* forces the current to flow in a smaller area of the conductor.

Resistance changes with temperature as

$$R_{t2} = R_{t1} \frac{M + t_2}{M + t_1}$$

where

R_{t2} = resistance at temperature t_2 given in °C
R_{t1} = resistance at temperature t_1 given in °C
M = a temperature coefficient for the given material
= 228.1 for aluminum
= 241.5 for annealed hard-drawn copper

For a wide range of temperatures, resistance rises almost linearly with temperature for both aluminum and copper. The effect of temperature is simplified as a linear equation as

$$R_{t2} = R_{t1}[1 + \alpha(t_2 - t_1)]$$

where

α = a temperature coefficient of resistance
= 0.00404 for 61.2% IACS aluminum at 20°C
= 0.00347 for 6201-T81 aluminum alloy at 20°C
= 0.00383 for hard-drawn copper at 20°C
= 0.0036 for aluminum-clad steel at 20°C

So, the resistance of aluminum with a 61.2% conductivity rises 4% for every 10°C rise in temperature.

We can also linearly interpolate using resistances provided at two different temperatures as

$$R(T_c) = R(T_{low}) + \frac{R(T_{high}) - R(T_{low})}{T_{high} - T_{low}}(T_c - T_{low})$$

where

$R(T_c)$ = conductor resistance at temperature T_c
$R(T_{high})$ = resistance at the higher temperature T_{high}
$R(T_{low})$ = resistance at the lower temperature T_{low}

With alternating current, skin effects raise the resistance of a conductor relative to its dc resistance. At 60 Hz, the resistance of a conductor is very close to its dc resistance except for very large conductors. Skin effects are much more important for high-frequency analysis such as switching surges and power-line carrier problems. They play a larger role in larger conductors.

The internal resistance of a solid round conductor including skin effects is [for details, see Stevenson (1962)]:

$$\frac{R_{ac}}{R_{dc}} = \frac{x}{2} \frac{\text{ber}(x)\text{bei}'(x) - \text{bei}(x)\text{ber}'(x)}{(\text{bei}'(x))^2 + (\text{ber}'(x))^2}$$

where

$$x = 0.02768 \sqrt{\frac{f\mu}{R_{dc}}}$$

f = frequency in Hz
μ = relative permeability = 1 for nonmagnetic conductors (including aluminum and copper)
R_{dc} = dc resistance of the conductor in ohms/1000 ft

ber, bei, ber', and bei' = real and imaginary modified Bessel functions and their derivatives (also called Kelvin functions)

For x greater than 3 (frequencies in the kilohertz range), the resistance increases linearly with x (Clarke, 1950) approximately as

$$\frac{R_{ac}}{R_{dc}} = \frac{x}{2\sqrt{2}} + \frac{1}{4} = 0.009786 \sqrt{\frac{f\mu}{R_{dc}}} + 0.25$$

So, for higher frequencies, the ac resistance increases as the square root of the frequency. For most distribution power-frequency applications, we can ignore skin effects (and they are included in ac resistance tables).

For most cases, we can model a stranded conductor as a solid conductor with the same cross-sectional area. ACSR with a steel core is slightly different. Just as in a transformer, the steel center conductor has losses due to hysteresis and eddy currents. If an ACSR conductor has an even number of layers, the axial magnetic field produced by one layer tends to cancel that produced by the next layer. We can model these as a tubular conductor for calculating skin effect. For odd numbers of layers, especially single-layered conductors like 6/1 or 7/1, the 60-Hz/dc ratio is higher than normal, especially at high current densities. These effects are reflected in the resistances included in tables (such as Table 2.3).

The reactance part of the impedance usually dominates the impedances on overhead circuit for larger conductors; below 4/0, resistance plays more of a role. For all-aluminum conductors on a 10-ft crossarm, the resistance approximately equals the reactance for a 2/0 conductor. Reactance is proportional to inductance; and inductance causes a voltage that opposes the change in the flow of current. Alternating current is always changing, so a reactance always creates a voltage due to current flow.

Distance between conductors determines the external component of reactance. Inductance is based on the area enclosed by a loop of current; a larger area (more separation between conductors) has more inductance. On overhead circuits, reactance of the line is primarily based on the separations between conductors — not the size of the conductor, not the type of metal in the conductor, not the stranding of the conductor.

The reactance between two parallel conductors in ohms per mile is:

$$X_{ab} = 0.2794 \frac{f}{60} \log_{10} \frac{d_{ab}}{GMR}$$

where

f = frequency in hertz
d_{ab} = distance between the conductors
GMR = geometric mean radius of both conductors

d_{ab} and GMR must have the same units, normally feet. More separation — a bigger loop — gives larger impedances.

The geometric mean radius (GMR) quantifies a conductor's internal inductance — by definition, the GMR is the radius of an infinitely thin tube having the same internal inductance as the conductor out to a one-foot radius. The GMR is normally given in feet to ease calculations with distances measured in feet. GMR is less than the actual conductor radius. Many conductor tables provide x_a, the inductive reactance due to flux in the conductor and outside the conductor out to a one-foot radius. The GMR in feet at 60 Hz relates to x_a as:

$$x_a = 0.2794 \log_{10} \frac{1}{GMR}$$

where GMR is in feet, and x_a is in ohms/mile.

For a solid, round, nonmagnetic conductor, the relationship between the actual radius and the GMR is

$$\frac{GMR}{r} = e^{-1/4} = 0.779$$

For stranded conductors, the GMR is

$$GMR = k \cdot r$$

TABLE 2.6

GMR Factor

Strands	GMR Factor, k
1 (solid)	0.7788
3	0.6778
7	0.7256
19	0.7577
37	0.7678
61	0.7722

Source: Aluminum Association, *Ampacities for Aluminum and ACSR Overhead Electrical Conductors*, 1986.

where

k = the *GMR* factor from Table 2.6
r = conductor radius

For ACSR conductors (which are layered), the *GMR* factor is more complicated.

Current flowing in a conductor induces a reactive voltage drop on the conductor it is flowing through. By the same induction, current flow in one conductor creates a voltage gradient along parallel conductors (see Figure 2.3). This voltage is of the same polarity as the voltage on the current-carrying conductor. Closer conductors have larger induced voltages. This induction is significant for several reasons:

- *Opposite flow* — Current flows more easily when a parallel conductor has flow in the opposite direction. The magnetic field from the other conductor creates a voltage drop that encourages flow in the opposite direction. Conductors carrying current in opposite directions have lower impedance when they are closer together.

- *Parallel flow* — A conductor carrying current in the same direction as a parallel conductor faces more impedance because of the current in the other conductor. Conductors carrying current in the same direction have higher impedance when they are closer together.

- *Circulating current* — Current flow in the vicinity of a shorted current loop induces currents to circulate in the loop.

For balanced conditions — balanced voltages, balanced loads, and balanced impedances — we can analyze power systems just with positive-sequence voltages, currents, and impedances. This is regularly done in transmission-planning and industrial load-flow studies. Using just positive-sequence quantities simplifies analysis; it's like a single-phase circuit rather than a three-phase circuit. For distribution circuits, unbalanced loading is quite common, so we normally need more than just positive-sequence parameters — we need the zero-sequence parameters as well. We also need unbalanced analysis approaches for phase-to-ground or phase-to-phase faults.

Mutual induction:

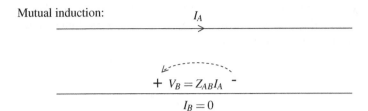

I_A

$+ \ V_B = Z_{AB}I_A \ -$

$I_B = 0$

Effects of induction:

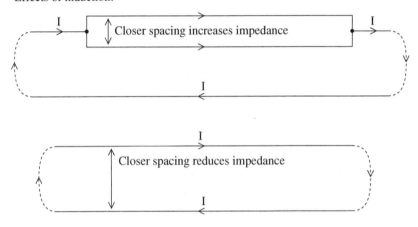

Closer spacing increases impedance

Closer spacing reduces impedance

FIGURE 2.3
Mutual induction.

With symmetrical components, the phasors of circuit quantities on each of the three phases resolve to three sets of phasors. For voltage, the symmetrical components relate to the phase voltages as:

$$V_a = V_0 + V_1 + V_2 \qquad V_0 = 1/3 \ (V_a + V_b + V_c)$$

$$V_b = V_0 + a^2V_1 + aV_2 \quad \text{and} \quad V_1 = 1/3 \ (V_a + aV_b + a^2V_c)$$

$$V_c = V_0 + aV_1 + a^2V_2 \qquad V_2 = 1/3 \ (V_a + a^2V_b + aV_c)$$

where $a = 1\angle 120°$ and $a^2 = 1\angle 240°$.

These phase-to-symmetrical conversions apply for phase-to-ground as well as phase-to-phase voltages. The same conversions apply for converting line currents to sequence currents:

$$I_a = I_0 + I_1 + I_2 \qquad I_0 = 1/3 \ (I_a + I_b + I_c)$$

$$I_b = I_0 + a^2I_1 + aI_2 \quad \text{and} \quad I_1 = 1/3 \ (I_a + aI_b + a^2I_c)$$

$$I_c = I_0 + aI_1 + a^2I_2 \qquad I_2 = 1/3 \ (I_a + a^2I_b + aI_c)$$

The voltage drop along each of the phase conductors depends on the currents in each of the phase conductors and the self impedances (such as Z_{aa}) and the mutual impedances (such as Z_{ab}) as

$$V_a = Z_{aa}I_a + Z_{ab}I_b + Z_{ac}I_c$$

$$V_b = Z_{ba}I_a + Z_{bb}I_b + Z_{bc}I_c$$

$$V_c = Z_{ca}I_a + Z_{cb}I_b + Z_{cc}I_c$$

Likewise, when we use sequence components, we have voltage drops of each sequence in terms of the sequence currents and sequence impedances:

$$V_0 = Z_{00}I_0 + Z_{01}I_1 + Z_{02}I_2$$

$$V_1 = Z_{10}I_0 + Z_{11}I_1 + Z_{12}I_2$$

$$V_2 = Z_{20}I_0 + Z_{21}I_1 + Z_{22}I_2$$

This is not much of a simplification until we assume that all of the self-impedance terms are equal ($Z_S = Z_{aa} = Z_{bb} = Z_{cc}$) and all of the mutual impedances are equal ($Z_M = Z_{ab} = Z_{ac} = Z_{bc} = Z_{ba} = Z_{ca} = Z_{cb}$). With this assumption, the sequence impedances decouple; the mutual terms of the zero-sequence matrix (such as Z_{12}) become zero. Zero-sequence current only causes a zero-sequence voltage drop. This is a good enough approximation for many distribution problems and greatly simplifies hand and computer calculations. Now, the sequence voltage drop equations are:

$$V_0 = Z_{00}I_0 = (Z_S + 2Z_M)I_0$$

$$V_1 = Z_{11}I_1 = (Z_S - Z_M)I_1$$

$$V_2 = Z_{22}I_2 = (Z_S - Z_M)I_2$$

Now, we have the sequence terms as

$$Z_0 = Z_S + 2Z_M$$

$$Z_1 = Z_2 = Z_S - Z_M$$

And likewise,

$$Z_S = (Z_0 + 2Z_1)/3$$

$$Z_M = (Z_0 - Z_1)/3$$

Note Z_S, the self-impedance term. Z_S is also the "loop impedance" — the impedance to current through one phase wire and returning through the ground return path. This loop impedance is important because it is the impedance for single-phase lines and the impedance for single line-to-ground faults.

Engineers normally use three methods to find impedances of circuits. In order of least to most accurate, these are:

- Table lookup
- Hand calculations
- Computer calculations

This book provides data necessary for the first two approaches. Table lookups are quite common. Even though table lookup is not the most accurate approach, its accuracy is good enough for analyzing most common distribution problems. Computer calculations are quite accessible and allow easier analysis of more complicated problems.

2.4 Simplified Line Impedance Calculations

The positive-sequence impedance of overhead lines is

$$Z_1 = R_\phi + jk_1 \log_{10} \frac{GMD_\phi}{GMR_\phi}$$

where

R_ϕ = resistance of the phase conductor in Ω/distance

k_1 = 0.2794f/60 for outputs in Ω/mi

  = 0.0529f/60 for outputs in Ω/1000 ft

f = frequency in hertz

GMR_ϕ = geometric mean radius of the phase conductor in ft

GMD_ϕ = geometric mean distance between the phase conductors in ft

GMD_ϕ = $\sqrt[3]{d_{AB}d_{BC}d_{CA}}$ for three-phase lines

GMD_ϕ = 1.26 d_{AB} for a three-phase line with flat configuration, either horizontal or vertical, where $d_{AB} = d_{BC} = 0.5d_{CA}$

GMD_ϕ = d_{AB} for two-phase lines*

---

* The two-phase circuit has two out of the three phases; the single-phase circuit has one phase conductor with a neutral return. While it may seem odd to look at the positive-sequence impedance of a one- or two-phase circuit, the analysis approach is useful. This approach uses fictitious conductors for the missing phases to model the one- or two-phase circuit as an equivalent three-phase circuit (no current flows on these fictitious phases).

$GMD_\phi = d_{AN}$ for single-phase lines*
d_{ij}　　= distance between the center of conductor i and the center of conductor j, in feet

For 60 Hz and output units of $\Omega/1000$ ft, this is

$$Z_1 = R_\phi + j0.0529 \log_{10} \frac{GMD_\phi}{GMR_\phi}$$

Zero-sequence impedance calculations are more complicated than positive-sequence calculations. Carson's equations are the most common way to account for the ground return path in impedance calculations of overhead circuits. Carson (1926) derived an expression including the earth return path. We'll use a simplification of Carson's equations; it includes the following assumptions (Smith, 1980)

- Since distribution lines are relatively short, the height-dependent terms in Carson's full model are small, so we neglect them.
- The multigrounded neutral is perfectly coupled to the earth (this has some drawbacks for certain calculations as discussed in Chapter 13).
- End effects are neglected.
- The current at the sending end equals that at the receiving end (no leakage currents).
- All phase conductors have the same size conductor.
- The ground is infinite and has uniform resistivity.

Consider the circuit in Figure 2.4; current flows in conductor a and returns through the earth. The voltage on conductor a equals the current times Z_{aa}, which is the self-impedance with an earth return path. The current in conductor a induces a voltage drop along conductor b equaling the phase-a current times Z_{ab}, which is the mutual impedance with an earth return path. These two impedances are found (Smith, 1980) with

$$Z_{aa} = R_\phi + R_e + jk_1 \log_{10} \frac{D_e}{GMR_\phi}$$

$$Z_{ab} = R_e + jk_1 \log_{10} \frac{D_e}{d_{ab}}$$

where
　R_e = resistance of the earth return path
　　= $0.0954f/60 \ \Omega/\text{mi}$
　　= $0.01807f/60 \ \Omega/1000$ ft

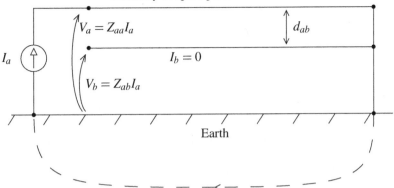

FIGURE 2.4
Depiction of Carson's impedances with earth return.

$D_e = 2160\sqrt{\rho/f}$ = equivalent depth of the earth return current in ft
ρ = earth resistivity in Ω-m
d_{ab} = distance between the centers of conductors a and b

For 60 Hz and output units of $\Omega/1000$ ft,

$$Z_{aa} = R_\phi + 0.01807 + j0.0529\log_{10}\frac{278.9\sqrt{\rho}}{GMR_\phi}$$

$$Z_{ab} = 0.01807 + j0.0529\log_{10}\frac{278.9\sqrt{\rho}}{d_{ab}}$$

These equations lead to different formulations for the zero-sequence impedance of circuits depending on the grounding configuration. They are also useful in their own right in many circumstances. Single-phase circuits with a phase and a neutral are often easier to analyze using these equations rather than using sequence components. Consider a single-phase circuit that is perfectly grounded with a current of I_A in the phase conductor. As Figure 2.5 shows, we can find the neutral current as a function of the mutual impedance between the two conductors divided by the self-impedance of the neutral conductor.

Now, let's look at the zero-sequence impedances — these change depending on grounding configuration. Figure 2.6 shows the configurations that we will consider.

A three-wire overhead line has a zero-sequence impedance of (Smith, 1980):

$$Z_0 = R_\phi + 3R_e + j3k_1\log_{10}\frac{D_e}{\sqrt[3]{GMR_\phi \cdot GMD_\phi^2}}$$

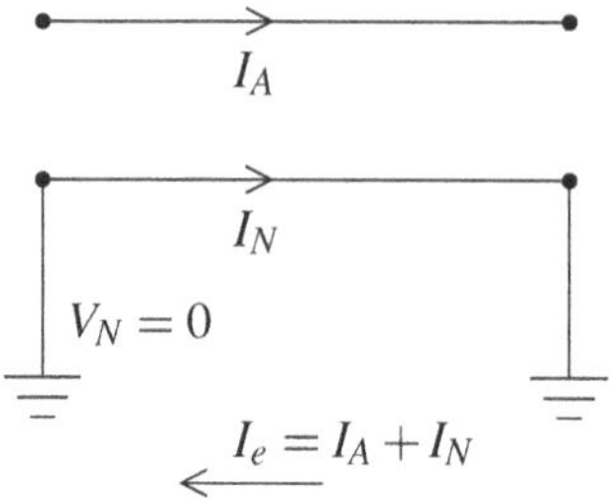

I_A

$V_N = 0 = Z_{AN}I_A + Z_{NN}I_N$

I_N

So, $I_N = -\dfrac{Z_{AN}}{Z_{NN}}I_A$

$V_N = 0$

$I_e = I_A + I_N$

FIGURE 2.5
Current flow in a neutral conductor based on self-impedances and mutual impedances.

For a four-wire multigrounded system, the zero-sequence self-impedance is:

$$Z_0 = R_\phi + 3R_e + j3k_1 \log_{10} \frac{D_e}{\sqrt[3]{GMR_\phi \cdot GMD_\phi^2}} - 3\frac{Z_{\phi N}^2}{Z_{NN}}$$

where Z_{NN} is the self-impedance of the neutral conductor with an earth return, and $Z_{\phi N}$ is the mutual impedance between the phase conductors as a group and the neutral. For 60 Hz and output units of $\Omega/1000$ ft, the zero-sequence self-impedance is

$$Z_0 = R_\phi + 0.0542 + j0.1587 \log_{10} \frac{278.9\sqrt{\rho}}{\sqrt[3]{GMR_\phi \cdot GMD_\phi^2}} - 3\frac{Z_{\phi N}^2}{Z_{NN}}$$

$$Z_{NN} = R_N + 0.01807 + j0.0529 \log_{10} \frac{278.9\sqrt{\rho}}{GMR_N}$$

$$Z_{\phi N} = 0.01807 + j0.0529 \log_{10} \frac{278.9\sqrt{\rho}}{GMD_{\phi N}}$$

where
    GMR_N = geometric mean radius of the neutral conductor in ft
    $GMD_{\phi N}$ = geometric mean distance between the phase conductors as a group and the neutral in ft
    $GMD_{\phi N} = \sqrt[3]{d_{AN}d_{BN}d_{CN}}$ for three-phase lines
    $GMD_{\phi N} = \sqrt{d_{AN}d_{BN}}$ for two-phase lines
    $GMD_{\phi N} = d_{AN}$ for single-phase lines

A special case is for a four-wire unigrounded circuit where the return current stays in the neutral, which has a zero-sequence impedance of (Ender et al., 1960)

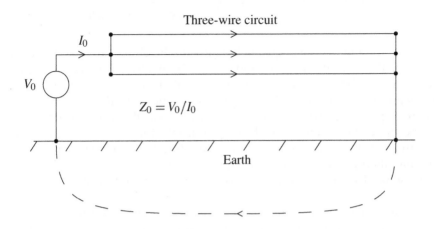

Three-wire circuit

I_0

V_0

$Z_0 = V_0/I_0$

Earth

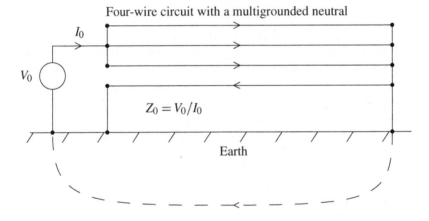

Four-wire circuit with a multigrounded neutral

I_0

V_0

$Z_0 = V_0/I_0$

Earth

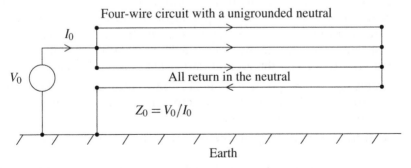

Four-wire circuit with a unigrounded neutral

I_0

V_0

All return in the neutral

$Z_0 = V_0/I_0$

Earth

FIGURE 2.6
Different zero-sequence impedances depending on the grounding configuration.

$$Z_0 = R_\phi + 3R_n + jR_n + j3k_1 \log_{10} \frac{GMD_{\phi N^2}}{GMR_N \sqrt[3]{GMR_\phi \cdot GMD_\phi^2}}$$

This is for a four-wire unigrounded circuit where there are no connections between the neutral conductor and earth. We can also use this as an approximation for a multigrounded neutral line that is very poorly grounded. Remember that the equation given above for a multigrounded circuit assumes perfect grounding. For some calculations, that is not accurate. This is the opposite extreme, which is appropriate for some calculations. Lat (1990) used this as one approach to estimating the worst-case overvoltage on unfaulted phases during a line-to-ground fault.

So, what does all of this mean? Some of the major effects are:

- *Conductor size* — Mainly affects resistance — larger conductors have lower positive-sequence resistance. Positive-sequence reactance also lowers with larger conductor size, but since it changes with the logarithm of conductor radius, changes are small.

- *Conductor spacings* — Increasing spacing (higher GMD_ϕ) increases Z_1. Increasing spacing reduces Z_0. Both of these changes with spacing are modest given the logarithmic effect.

- *Neutral* — Adding the neutral always reduces the total zero-sequence impedance, $|Z_0|$. Adding a neutral always reduces the reactive portion of this impedance. But adding a neutral may increase or may decrease the resistive portion of Z_0. Adding a small neutral with high resistance increases the resistance component of Z_0.

- *Neutral spacing* — Moving the neutral closer to the phase conductors reduces the zero-sequence impedance (but may increase the resistive portion, depending on the size of the neutral).

- *Earth resistivity* — The earth resistivity does not change the earth return resistance (R_e only depends on frequency). The current spreads to wider areas of the earth in high-resistivity soil. Earth resistivity does change the reactance. Higher earth resistivities force current deeper into the ground (larger D_e), raising the reactance.

- *Grounding* — The positive-sequence impedance Z_1 stays the same regardless of the grounding, whether four-wire multigrounded, ungrounded, or unigrounded.

- *Negative sequence* — Equals the positive-sequence impedance.

Figure 2.7 and Figure 2.8 show the effects of various parameters on the positive and zero-sequence impedances. Many of the outputs are not particularly sensitive to changes in the inputs. Since many parameters are functions of the logarithm of the variable, major changes induce only small changes in the impedance.

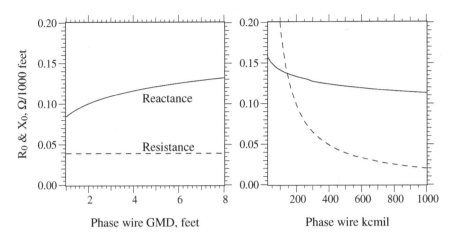

FIGURE 2.7
Effect of spacings and conductor size on the positive-sequence impedance with 500-kcmil AAC phases (GMR = 0.0256 ft) and GMD_ϕ=5 ft.

When do we need more accuracy or more sophistication? For power flows, fault calculations, voltage flicker calculations, and voltage sag analysis, we normally don't need more sophistication. For switching surges, lightning, or other higher frequency transient analysis, we normally need more sophisticated line models.

Most unbalanced calculations can be done with this approach, but some cases require more sophistication. Distribution lines and most lower-voltage subtransmission lines are not transposed. On some long circuits, even with balanced loading, the unbalanced impedances between phases creates voltage unbalance.

2.5 Line Impedance Tables

This section has several tables of impedances for all-aluminum, ACSR, and copper constructions. All are based on the equations in the previous section and assume GMD = 4.8 ft, conductor temperature = 50°C, $GMD_{\phi N}$ = 6.3 ft, and earth resistivity = 100 Ω-m. All zero-sequence values are for a four-wire multigrounded neutral circuit.

2.6 Conductor Sizing

We have an amazing variety of sizes and types of conductors. Several electrical, mechanical, and economic characteristics affect conductor selection:

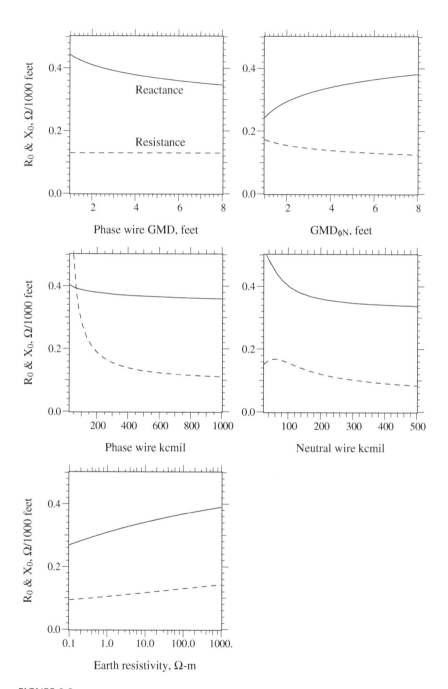

FIGURE 2.8
Effect of various parameters on the zero-sequence impedance with a base case of AAC 500-kcmil phases, 3/0 neutral (168 kcmil), GMD_ϕ = 5 ft, $GMD_{\phi N}$ = 6.3 ft, and ρ = 100 Ω-m.

TABLE 2.7

Positive-Sequence Impedances of All-Aluminum Conductor

Phase Size	Strands	R_1	X_1	Z_1
6	7	0.7405	0.1553	0.7566
4	7	0.4654	0.1500	0.4890
2	7	0.2923	0.1447	0.3262
1	7	0.2323	0.1420	0.2723
1/0	7	0.1839	0.1394	0.2308
2/0	7	0.1460	0.1367	0.2000
3/0	7	0.1159	0.1341	0.1772
4/0	7	0.0920	0.1314	0.1604
250	7	0.0778	0.1293	0.1509
266.8	7	0.0730	0.1286	0.1478
300	19	0.0649	0.1261	0.1418
336.4	19	0.0580	0.1248	0.1376
350	19	0.0557	0.1242	0.1361
397.5	19	0.0490	0.1229	0.1323
450	19	0.0434	0.1214	0.1289
477	19	0.0411	0.1208	0.1276
500	19	0.0392	0.1202	0.1265
556.5	19	0.0352	0.1189	0.1240
700	37	0.0282	0.1159	0.1192
715.5	37	0.0277	0.1157	0.1190
750	37	0.0265	0.1151	0.1181
795	37	0.0250	0.1146	0.1173
874.5	37	0.0227	0.1134	0.1157
900	37	0.0221	0.1130	0.1152
954	37	0.0211	0.1123	0.1142
1000	37	0.0201	0.1119	0.1137

Note: Impedances, $\Omega/1000$ ft ($\times$ 5.28 for $\Omega/$mi or $\times$ 3.28 for $\Omega/$km). GMD = 4.8 ft, Conductor temp. = 50°C.

- *Ampacity* — The peak current-carrying capability of a conductor limits the current (and power) carrying capability.

- *Economics* — Often we will use a conductor that normally operates well below its ampacity rating. The cost of the extra aluminum pays for itself with lower I^2R losses; the conductor runs cooler. This also leaves room for expansion.

- *Mechanical strength* — Especially on rural lines with long span lengths, mechanical strength plays an important role in size and type of conductor. Stronger conductors like ACSR are used more often. Ice and wind loadings must be considered.

- *Corrosion* — While not usually a problem, corrosion sometimes limits certain types of conductors in certain applications.

As with many aspects of distribution operations, many utilities standardize on a set of conductors. For example, a utility may use 500-kcmil AAC

TABLE 2.8

AAC Zero-Sequence, Z_0, and Ground-Return Loop Impedances, $Z_S = (2Z_1 + Z_0)/3$

Phase Size	Neutral Size	R_0	X_0	Z_0	R_S	X_S	Z_S
6	6	0.8536	0.5507	1.0158	0.7782	0.2871	0.8294
2	6	0.4055	0.5401	0.6754	0.3301	0.2765	0.4306
2	2	0.4213	0.4646	0.6272	0.3353	0.2513	0.4190
1	6	0.3454	0.5374	0.6389	0.2700	0.2738	0.3845
1	1	0.3558	0.4405	0.5662	0.2734	0.2415	0.3648
1/0	6	0.2971	0.5348	0.6117	0.2216	0.2712	0.3502
1/0	1/0	0.2981	0.4183	0.5136	0.2220	0.2323	0.3213
2/0	6	0.2591	0.5321	0.5919	0.1837	0.2685	0.3253
2/0	2/0	0.2487	0.3994	0.4705	0.1802	0.2243	0.2877
3/0	2	0.2449	0.4540	0.5158	0.1589	0.2407	0.2884
3/0	3/0	0.2063	0.3840	0.4359	0.1461	0.2174	0.2619
4/0	1	0.2154	0.4299	0.4809	0.1331	0.2309	0.2665
4/0	4/0	0.1702	0.3716	0.4088	0.1180	0.2115	0.2422
250	2/0	0.1805	0.3920	0.4316	0.1120	0.2169	0.2441
250	250	0.1479	0.3640	0.3929	0.1012	0.2075	0.2309
266.8	2/0	0.1757	0.3913	0.4289	0.1072	0.2161	0.2413
266.8	266.8	0.1402	0.3614	0.3877	0.0954	0.2062	0.2272
300	2/0	0.1675	0.3888	0.4234	0.0991	0.2137	0.2355
300	300	0.1272	0.3552	0.3773	0.0856	0.2025	0.2198
336.4	2/0	0.1607	0.3875	0.4195	0.0922	0.2123	0.2315
336.4	336.4	0.1157	0.3514	0.3699	0.0772	0.2003	0.2147
350	2/0	0.1584	0.3869	0.4181	0.0899	0.2118	0.2301
350	350	0.1119	0.3499	0.3674	0.0744	0.1994	0.2129
397.5	2/0	0.1517	0.3856	0.4143	0.0832	0.2105	0.2263
397.5	397.5	0.1005	0.3463	0.3606	0.0662	0.1974	0.2082
450	2/0	0.1461	0.3841	0.4109	0.0776	0.2089	0.2229
450	450	0.0908	0.3427	0.3545	0.0592	0.1951	0.2039
477	2/0	0.1438	0.3835	0.4096	0.0753	0.2084	0.2216
477	477	0.0868	0.3414	0.3522	0.0563	0.1943	0.2023
500	4/0	0.1175	0.3605	0.3791	0.0653	0.2003	0.2107
500	500	0.0835	0.3401	0.3502	0.0540	0.1935	0.2009
556.5	4/0	0.1135	0.3591	0.3766	0.0613	0.1990	0.2082
556.5	556.5	0.0766	0.3372	0.3458	0.0490	0.1917	0.1978
700	4/0	0.1064	0.3561	0.3717	0.0542	0.1960	0.2033
700	700	0.0639	0.3310	0.3371	0.0401	0.1876	0.1918
715.5	4/0	0.1060	0.3559	0.3714	0.0538	0.1958	0.2030
715.5	715.5	0.0632	0.3306	0.3366	0.0395	0.1873	0.1915
750	4/0	0.1047	0.3553	0.3705	0.0526	0.1952	0.2022
750	750	0.0609	0.3295	0.3351	0.0380	0.1866	0.1904
795	4/0	0.1033	0.3548	0.3695	0.0511	0.1946	0.2012
795	795	0.0582	0.3283	0.3335	0.0361	0.1858	0.1893
874.5	4/0	0.1010	0.3536	0.3678	0.0488	0.1935	0.1996
874.5	874.5	0.0540	0.3261	0.3305	0.0332	0.1843	0.1873
900	4/0	0.1004	0.3533	0.3672	0.0482	0.1931	0.1990
900	900	0.0529	0.3254	0.3296	0.0324	0.1838	0.1866
954	4/0	0.0993	0.3525	0.3662	0.0471	0.1924	0.1981

TABLE 2.8 (Continued)

AAC Zero-Sequence, Z_0, and Ground-Return Loop Impedances, $Z_S = (2Z_1 + Z_0)/3$

Phase Size	Neutral Size	R_0	X_0	Z_0	R_S	X_S	Z_S
954	954	0.0510	0.3239	0.3279	0.0310	0.1828	0.1854
1000	4/0	0.0983	0.3521	0.3656	0.0462	0.1920	0.1975
1000	1000	0.0491	0.3232	0.3269	0.0298	0.1823	0.1847

Note: Impedances, Ω/1000 ft ($\times$ 5.28 for Ω/mi or $\times$ 3.28 for Ω/km). GMD = 4.8 ft, $GMD_{\phi N}$ = 6.3 ft, Conductor temp. = 50°C, Earth resistivity = 100 Ω-m.

TABLE 2.9

Positive-Sequence Impedances of ACSR

Phase Size	Strands	R_1	X_1	Z_1
6	6/1	0.7500	0.1746	0.7700
4	6/1	0.4794	0.1673	0.5077
2	6/1	0.3080	0.1596	0.3469
1	6/1	0.2474	0.1551	0.2920
1/0	6/1	0.1972	0.1496	0.2476
2/0	6/1	0.1616	0.1478	0.2190
3/0	6/1	0.1208	0.1442	0.1881
4/0	6/1	0.1066	0.1407	0.1765
266.8	18/1	0.0723	0.1262	0.1454
336.4	18/1	0.0574	0.1236	0.1362
397.5	18/1	0.0487	0.1217	0.1311
477	18/1	0.0405	0.1196	0.1262
556.5	18/1	0.0348	0.1178	0.1228
636	18/1	0.0306	0.1165	0.1204
795	36/1	0.0247	0.1140	0.1167

Note: Impedances, Ω/1000 ft ($\times$ 5.28 for Ω/mi or $\times$ 3.28 for Ω/km). GMD = 4.8 ft, Conductor temp. = 50°C.

for all mainline spans and 1/0 AAC for all laterals. While many circuit locations are overdesigned, the utility saves from reduced stocking, fewer tools, and standardized connectors. While many utilities have more than just two conductors, most use just a handful of standard conductors; four to six economically covers the needs of most utilities.

2.7 Ampacities

The ampacity is the maximum designed current of a conductor. This current carrying capacity is normally given in amperes. A given conductor has

TABLE 2.10

ACSR Zero-Sequence, Z_0, and Ground-Return Loop Impedances, $Z_S = (2Z_1 + Z_0)/3$

Phase Size	Neutral Size	R_0	X_0	Z_0	R_S	X_S	Z_S
4	4	0.6030	0.5319	0.8040	0.5206	0.2888	0.5953
2	4	0.4316	0.5242	0.6790	0.3492	0.2812	0.4483
2	2	0.4333	0.4853	0.6505	0.3498	0.2682	0.4407
1	4	0.3710	0.5197	0.6385	0.2886	0.2766	0.3998
1	1	0.3684	0.4610	0.5901	0.2877	0.2571	0.3858
1/0	4	0.3208	0.5143	0.6061	0.2384	0.2712	0.3611
1/0	1/0	0.3108	0.4364	0.5357	0.2351	0.2452	0.3397
2/0	2	0.2869	0.4734	0.5536	0.2034	0.2563	0.3272
2/0	2/0	0.2657	0.4205	0.4974	0.1963	0.2387	0.3090
3/0	2	0.2461	0.4698	0.5304	0.1626	0.2527	0.3005
3/0	3/0	0.2099	0.4008	0.4524	0.1505	0.2297	0.2746
4/0	1	0.2276	0.4465	0.5012	0.1469	0.2426	0.2836
4/0	4/0	0.1899	0.3907	0.4344	0.1344	0.2240	0.2612
266.8	2/0	0.1764	0.3990	0.4362	0.1070	0.2171	0.2421
266.8	266.8	0.1397	0.3573	0.3836	0.0948	0.2032	0.2242
336.4	2/0	0.1615	0.3963	0.4280	0.0921	0.2145	0.2334
336.4	336.4	0.1150	0.3492	0.3676	0.0766	0.1988	0.2130
397.5	2/0	0.1528	0.3944	0.4230	0.0834	0.2126	0.2284
397.5	397.5	0.1002	0.3441	0.3584	0.0659	0.1958	0.2066
477	2/0	0.1446	0.3923	0.4181	0.0752	0.2105	0.2235
477	477	0.0860	0.3391	0.3498	0.0557	0.1927	0.2006
556.5	2/0	0.1389	0.3906	0.4145	0.0695	0.2087	0.2200
556.5	556.5	0.0759	0.3351	0.3436	0.0485	0.1902	0.1963

Note: Impedances, $\Omega/1000$ ft ($\times 5.28$ for Ω/mi or $\times 3.28$ for Ω/km). GMD = 4.8 ft, $\text{GMD}_{\phi N} = 6.3$ ft, Conductor temp. = 50°C, Earth resistivity = 100 Ω-m.

TABLE 2.11

Positive-Sequence Impedances of Hard-Drawn Copper

Phase Size	Strands	R_1	X_1	Z_1
4	3	0.2875	0.1494	0.3240
2	3	0.1809	0.1441	0.2313
1	7	0.1449	0.1420	0.2029
1/0	7	0.1150	0.1393	0.1807
2/0	7	0.0911	0.1366	0.1642
3/0	12	0.0723	0.1316	0.1501
4/0	12	0.0574	0.1289	0.1411
250	12	0.0487	0.1270	0.1360
300	12	0.0407	0.1250	0.1314
350	19	0.0349	0.1243	0.1291
400	19	0.0307	0.1227	0.1265
450	19	0.0273	0.1214	0.1244
500	19	0.0247	0.1202	0.1227

Note: Impedances, $\Omega/1000$ ft ($\times 5.28$ for Ω/mi or $\times 3.28$ for Ω/km). GMD = 4.8 ft, Conductor temp. = 50°C.

TABLE 2.12

Copper Zero-Sequence, Z_0, and Ground-Return Loop Impedances, $Z_S = (2Z_1 + Z_0)/3$

Phase Size	Neutral Size	R_0	X_0	Z_0	R_S	X_S	Z_S
3	3	0.3515	0.4459	0.5678	0.2707	0.2468	0.3663
3	3	0.3515	0.4459	0.5678	0.2707	0.2468	0.3663
6	6	0.5830	0.5157	0.7784	0.4986	0.2751	0.5695
6	6	0.5830	0.5157	0.7784	0.4986	0.2751	0.5695
4	6	0.4141	0.5104	0.6572	0.3297	0.2697	0.4260
4	4	0.4146	0.4681	0.6253	0.3299	0.2556	0.4173
2	6	0.3075	0.5051	0.5913	0.2231	0.2644	0.3460
2	2	0.2924	0.4232	0.5143	0.2181	0.2371	0.3221
1	4	0.2720	0.4606	0.5349	0.1873	0.2482	0.3109
1	1	0.2451	0.4063	0.4745	0.1783	0.2301	0.2911
1/0	4	0.2421	0.4580	0.5180	0.1574	0.2455	0.2916
1/0	1/0	0.2030	0.3915	0.4410	0.1443	0.2234	0.2660
2/0	3	0.2123	0.4352	0.4842	0.1315	0.2362	0.2703
2/0	2/0	0.1672	0.3796	0.4148	0.1165	0.2176	0.2468
3/0	2	0.1838	0.4106	0.4498	0.1095	0.2246	0.2498
3/0	3/0	0.1383	0.3662	0.3915	0.0943	0.2098	0.2300
4/0	2	0.1689	0.4080	0.4415	0.0946	0.2219	0.2412
4/0	4/0	0.1139	0.3584	0.3760	0.0762	0.2054	0.2191
250	1	0.1489	0.3913	0.4187	0.0821	0.2151	0.2303
250	250	0.0993	0.3535	0.3671	0.0656	0.2025	0.2128
300	1	0.1409	0.3893	0.4140	0.0741	0.2131	0.2256
300	300	0.0856	0.3487	0.3590	0.0557	0.1995	0.2071
350	1	0.1351	0.3886	0.4115	0.0683	0.2124	0.2231
350	350	0.0754	0.3468	0.3549	0.0484	0.1985	0.2043
400	1	0.1309	0.3871	0.4086	0.0641	0.2109	0.2204
400	400	0.0680	0.3437	0.3503	0.0431	0.1964	0.2011
450	1/0	0.1153	0.3736	0.3910	0.0566	0.2055	0.2131
450	450	0.0620	0.3410	0.3466	0.0389	0.1946	0.1984
500	1/0	0.1127	0.3724	0.3891	0.0540	0.2043	0.2113
500	500	0.0573	0.3387	0.3435	0.0356	0.1930	0.1963

Note: Impedances, $\Omega/1000$ ft ($\times 5.28$ for $\Omega/$mi or $\times 3.28$ for $\Omega/$km). GMD = 4.8 ft, $\text{GMD}_{\phi N}$ = 6.3 ft, Conductor temp. = 50°C, Earth resistivity = 100 Ω-m.

several ampacities, depending on its application and the assumptions used. House and Tuttle (1958) derive the ampacity calculations described below, which are used in IEEE Std. 738-1993 and most other published ampacity tables (Aluminum Association, 1986; Southwire Company, 1994).

Sun, wind, and ambient temperature change a conductor's ampacity. A conductor's temperature depends on the thermal balance of heat inputs and losses. Current driven through a conductor's resistance creates heat (I^2R). The sun is another source of heat into the conductor. Heat escapes from the conductor through radiation and from convection. Considering the balance of inputs and outputs, the ampacity of a conductor is

$$I = \sqrt{\frac{q_c + q_r - q_s}{R_{ac}}}$$

where

q_c = convected heat loss, W/ft
q_r = radiated heat loss, W/ft
q_s = solar heat gain, W/ft
R_{ac} = Nominal ac resistance at operating temperature t, Ω/ft

The convected heat loss with no wind is

$$q_c = 0.283\sqrt{\rho_f}D^{0.75}(t_c - t_a)^{1.25}$$

Wind increases convection losses. The losses vary based on wind speed. The IEEE method uses the maximum q_c from the following two equations:

$$q_{c1} = \left[1.01 + 0.371\left(\frac{D\rho_f V}{\mu_f}\right)^{0.52}\right]K_f(t_c - t_a)$$

$$q_{c2} = 0.1695\left(\frac{D\rho_f V}{\mu_f}\right)^{0.6} K_f(t_c - t_a)$$

where

D = conductor diameter, in.
t_c = conductor operating temperature, °C
t_a = ambient temperature, °C
$t_f = (t_c + t_a)/2$
V = air velocity, ft/h
ρ_f = air density at t_f, lb/ft^3
μ_f = absolute viscosity of air at t_f, lb/h-ft
K_f = thermal conductivity of air at t_f, W/ft^2/°C

The density, viscosity, and thermal conductivity of air all depend on temperature (actually the film temperature T_f near the surface of the conductor, which is taken as the average of the conductor and ambient temperatures). Tables of these are available in the references (IEEE Std. 738-1993; Southwire Company, 1994). We may also use the following polynomial approximations (IEEE Std. 738-1993):

$$\rho_f = \frac{0.080695 - (0.2901 \times 10^{-5})H_c + (0.37 \times 10^{-10})H_c^2}{1 + 0.00367T_f}$$

where H_c is the altitude above sea level in feet.

$$k_f = 0.007388 + 2.27889 \times 10^{-5} T_f - 1.34328 \times 10^{-9} T_f^2$$

$$\mu_f = 0.0415 + 1.2034 \times 10^{-4} T_f - 1.1442 \times 10^{-7} T_f^2 + 1.9416 \times 10^{-10} T_f^3$$

A conductor radiates heat as the absolute temperature to the fourth power as

$$q_r = 0.138 D\varepsilon \left[\left(\frac{T_c + 273}{100} \right)^4 - \left(\frac{T_a + 273}{100} \right)^4 \right]$$

where
D = conductor diameter, in.
ε = emissivity (normally 0.23 to 0.91 for bare wires)
T_c = conductor temperature, °C
T_a = ambient temperature, °C

A conductor absorbs heat from the sun as

$$q_s = \alpha Q_s \frac{D}{12} \sin \theta$$

where
α = solar absorptivity
Q_s = total solar heat in W/ft^2
θ = effective angle of incidence of the sun's rays
D = conductor diameter, in.

The angles and total solar heat depend on the time of day and the latitude. Since the solar input term does not change the output significantly, we can use some default values. For a latitude of 30°N at 11 a.m. in clear atmosphere, Q_s = 95.2 W/ft,^{2} and θ = 78.6°.

Emissivity (ε) is the ability of a conductor to radiate heat into the air. Absorptivity (α) quantifies how much heat a conductor can absorb. Emissivity and absorptivity are interrelated; a nice shiny conductor reflects away much of the sun's heat but does not radiate heat well. Commonly, both are assumed to be 0.5 for bare wire. More conservative assumptions, possibly overconservative, are 0.7 for emissivity and 0.9 for absorptivity.

Some of the main factors impacting ampacity are

- *Allowable conductor temperature* — Ampacity increases significantly with higher allowed temperatures.

TABLE 2.13

Ampacities of All-Aluminum Conductor

Conductor	Stranding	Conductor Temp. = 75°C				Conductor Temp. = 100°C			
		Ambient = 25°C		Ambient = 40°C		Ambient = 25°C		Ambient = 40°C	
		No Wind	Wind	No Wind	Wind	No Wind	Wind	No Wind	Wind
6	7	60	103	46	85	77	124	67	111
4	7	83	138	63	114	107	166	92	148
2	7	114	185	86	152	148	223	128	199
1	7	134	214	101	175	174	258	150	230
1/0	7	157	247	118	203	204	299	176	266
2/0	7	184	286	139	234	240	347	207	309
3/0	7	216	331	162	271	283	402	243	358
4/0	7	254	383	190	313	332	466	286	414
250	7	285	425	213	347	373	518	321	460
250	19	286	427	214	348	375	519	322	462
266.8	7	298	443	223	361	390	539	335	479
266.8	19	299	444	224	362	392	541	337	481
300	19	325	479	243	390	426	584	367	519
336.4	19	351	515	262	419	461	628	397	559
350	19	361	527	269	428	474	644	408	572
397.5	19	394	571	293	464	517	697	445	619
450	19	429	617	319	501	564	755	485	671
477	19	447	640	332	519	588	784	506	697
477	37	447	641	333	520	589	785	507	697
500	19	461	658	342	534	606	805	521	716
556.5	19	496	704	368	571	654	864	562	767
556.5	37	496	705	369	571	654	864	563	768
600	37	522	738	388	598	688	905	592	804
636	37	545	767	404	621	720	943	619	838
650	37	556	782	413	633	737	965	634	857
700	37	581	814	431	658	767	1000	660	888
715.5	37	590	825	437	667	779	1014	670	901
715.5	61	590	825	437	667	780	1014	671	901
750	37	609	848	451	686	804	1044	692	927
795	37	634	881	470	712	840	1086	722	964
795	61	635	882	470	713	840	1087	723	965
800	37	636	884	471	714	841	1087	723	965
874.5	61	676	933	500	754	896	1152	770	1023
874.5	61	676	934	500	754	896	1152	771	1023
900	37	689	950	510	767	913	1172	785	1041
954	37	715	983	529	793	946	1210	813	1074
954	61	719	988	532	797	954	1221	821	1084
1000.0	37	740	1014	547	818	981	1252	844	1111

- *Ambient temperature* — Ampacity increases about 1% for each 1°C decrease in ambient temperature.
- *Wind speed* — Even a small wind helps cool conductors significantly. With no wind, ampacities are significantly lower than with a 2-ft/ sec crosswind.

Table 2.13 through Table 2.15 show ampacities of all-aluminum, ACSR, and copper conductors. All assume the following:

TABLE 2.14

Ampacities of ACSR

Conductor	Stranding	Conductor Temp. = 75°C				Conductor Temp. = 100°C			
		Ambient = 25°C		Ambient = 40°C		Ambient = 25°C		Ambient = 40°C	
		No Wind	Wind	No Wind	Wind	No Wind	Wind	No Wind	Wind
6	6/1	61	105	47	86	79	126	68	112
4	6/1	84	139	63	114	109	167	94	149
4	7/1	85	141	64	116	109	168	94	149
2	6/1	114	184	86	151	148	222	128	197
2	7/1	117	187	88	153	150	224	129	199
1	6/1	133	211	100	173	173	255	149	227
1/0	6/1	156	243	117	199	202	294	174	261
2/0	6/1	180	277	135	227	235	337	203	300
3/0	6/1	208	315	156	258	262	370	226	329
4/0	6/1	243	363	182	296	319	443	274	394
266.8	18/1	303	449	227	366	398	547	342	487
266.8	26/7	312	458	233	373	409	559	352	497
336.4	18/1	356	520	266	423	468	635	403	564
336.4	26/7	365	530	272	430	480	647	413	575
336.4	30/7	371	536	276	435	487	655	419	582
397.5	18/1	400	578	298	469	527	708	453	629
397.5	26/7	409	588	305	477	538	719	463	639
477	18/1	453	648	337	525	597	793	513	705
477	24/7	461	656	343	532	607	804	523	714
477	26/7	464	659	345	534	611	808	526	718
477	30/7	471	667	350	540	615	810	529	720
556.5	18/1	504	713	374	578	664	874	571	777
556.5	24/7	513	722	380	585	677	887	582	788
556.5	26/7	517	727	383	588	682	893	587	793
636	24/7	562	785	417	635	739	962	636	854
636	26/7	567	791	420	639	748	972	644	863
795	45/7	645	893	478	721	855	1101	735	977
795	26/7	661	910	489	734	875	1122	753	996
954	45/7	732	1001	541	807	971	1238	835	1099
954	54/7	741	1010	547	814	983	1250	846	1109
1033.5	45/7	769	1048	568	844	1019	1294	877	1148

- Emissivity = 0.5, absorptivity = 0.5
- 30°N at 11 a.m. in clear atmosphere
- Wind speed = 2 ft/sec
- Elevation = sea level

The solar heating input has modest impacts on the results. With no sun, the ampacity increases only a few percent.

Some simplifying equations help for evaluating some of the significant impacts on ampacity. We can estimate changes in ambient and allowable temperature variations (Black and Rehberg, 1985; Southwire Company, 1994) with

$$I_{new} = I_{old} \sqrt{\frac{T_{c,new} - T_{a,new}}{T_{c,old} - T_{a,old}}}$$

TABLE 2.15

Ampacities of Copper Conductors

		Conductor Temp. = 75°C				Conductor Temp. = 100°C			
		Ambient = 25°C		Ambient = 40°C		Ambient = 25°C		Ambient = 40°C	
Conductor	Stranding	No Wind	Wind	No Wind	Wind	No Wind	Wind	No Wind	Wind
6	3	83	140	63	116	107	169	92	150
6	1	76	134	58	110	98	160	85	143
5	3	97	162	73	134	125	195	108	174
5	1	90	155	68	127	115	185	99	165
4	3	114	188	86	154	147	226	127	201
4	1	105	179	80	147	136	214	117	191
3	7	128	211	97	174	166	254	143	226
3	3	133	217	101	178	173	262	149	233
3	1	123	206	93	170	159	248	137	221
2	7	150	244	114	201	195	294	168	262
2	3	157	251	118	206	203	303	175	270
2	1	145	239	110	196	187	287	161	256
1	3	184	291	138	238	239	351	206	313
1	7	177	282	133	232	229	340	197	303
1/0	7	207	326	156	267	269	394	232	351
2/0	7	243	378	183	309	317	457	273	407
3/0	12	292	444	219	362	381	539	328	479
3/0	7	285	437	214	357	373	530	321	472
4/0	19	337	507	252	414	440	617	379	549
4/0	12	342	513	256	418	448	624	386	555
4/0	7	335	506	251	413	438	615	377	547
250	19	377	563	282	459	493	684	424	608
250	12	384	569	287	464	502	692	432	615
300	19	427	630	319	513	559	767	481	682
300	12	435	637	324	519	569	776	490	690
350	19	475	694	355	565	624	847	537	753
350	12	484	702	360	571	635	858	546	763
400	19	520	753	387	612	682	920	587	817
450	19	564	811	420	659	742	993	639	883
500	37	606	865	450	702	798	1061	686	942
500	19	605	865	450	701	797	1059	685	941
600	37	685	968	509	784	905	1190	779	1057
700	37	759	1062	563	860	1003	1308	863	1161
750	37	794	1107	588	895	1051	1364	904	1211
800	37	826	1147	612	927	1092	1412	939	1253
900	37	894	1233	662	995	1189	1527	1023	1356
1000	37	973	1333	719	1075	1313	1676	1129	1488

where I_{new} is the new ampacity based on a new conductor limit $T_{c,new}$ and a new ambient temperature $T_{a,new}$. Likewise, I_{old} is the original ampacity based on a conductor limit $T_{c,old}$ and an ambient temperature $T_{a,old}$.

This approach neglects solar heating and the change in conductor resistance with temperature (both have small impacts). Doing this simplifies the ampacity calculation to a constant (dependent on weather and conductor characteristics) times the difference between the conductor temperature and the ambient temperature: $I^2 = K(T_c - T_a)$. We do not use this simplification for the original ampacity calculation, but it helps us evaluate changes in temperatures or currents.

We use this approach in Figure 2.9 to show the variation in ampacity with ambient conductor assumptions along with two conductor operating limits.

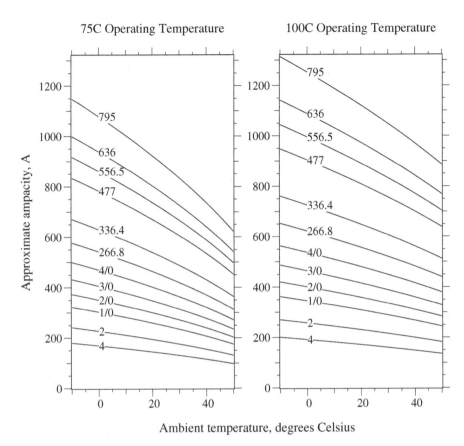

FIGURE 2.9
AAC ampacity with ambient temperature variations, using adjustments from base ampacity data in Table 2.13 (2 ft/sec wind, with sun).

Also, Figure 2.10 shows the conductor temperature vs. loading for several AAC conductors. This graph highlights the major impact of operating temperature on ampacity. If we are overly conservative on a conductor limit, we end up with an overly restrictive ampacity.

We can also use the simplified ampacity equation to estimate the conductor temperature at a current higher or lower than the rated ampacity as (and at a different ambient temperature if we wish):

$$T_{c,new} = T_{a,new} + \left(\frac{I_{new}}{I_{old}}\right)^2 (T_{c,old} - T_{a,old})$$

When examining a line's ampacity, always remember that the overhead wire may not be the weakest link; substation exit cables, terminations, reclosers, or other gear may limit a circuit's current before the conductors do. Also, with currents near a conductor's rating, voltage drop is high.

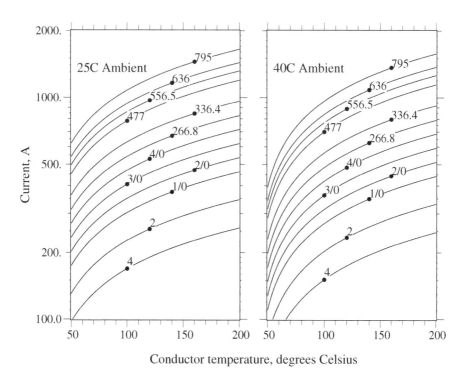

Conductor temperature, degrees Celsius

FIGURE 2.10
Conductor temperatures based on the given currents for selected AAC conductors, using adjustments from base ampacity data in Table 2.13 (2 ft/sec wind).

The maximum operating temperature is an important consideration. Higher designed operating temperatures allow higher currents. But at higher temperatures, we have a higher risk of damage to the conductors. Aluminum strands are strain hardened during manufacturing (the H19 in aluminum's 1350-H19 designation means "extra hard"). Heating relaxes the strands — the aluminum elongates and weakens. This damage is called *annealing*. As aluminum anneals, it reverts back to its natural, softer state: fully annealed 1350 aluminum wire elongates by 30% and loses 58% of its strength (10,000 psi vs. 24,000 psi fully hardened). Even fully annealed, failure may not be immediate; the next ice load or heavy winds may break a conductor. Slow annealing begins near 100°C. Aluminum anneals rapidly above 200°C. Annealing damage is permanent and accumulates over time. Remaining strength for AAC conductors varies with conductor temperature and duration of exposure as approximately (Harvey, 1971)

$$R_S = k_1 t^{-\frac{0.1}{d}(0.001T_c - 0.095)}$$

where
R_S = remaining strength, percent of initial strength

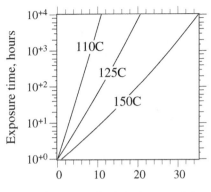

Percent of original strength lost

FIGURE 2.11
Loss of strength of all-aluminum conductors due to exposure to high temperatures.

d = strand diameter, in.
t = exposure time, h
T_c = conductor temperature, °C
$k_1 = (-0.24T_c + 135)$, but if $k_1 > 100$, use $k_1 = 100$

Figure 2.11 shows the loss of strength with time for high-temperature operation using this approximation.

ACSR may be loaded higher than the same size AAC conductor. As the aluminum loses strength, the steel carries more of the tension. The steel does not lose strength until reaching higher temperatures.

Covered conductors are darker, so they absorb more heat from the sun but radiate heat better; the Aluminum Association (1986) uses 0.91 for both the emissivity and the absorptivity of covered wire. Table 2.16 shows ampacities of covered wire. Covered conductors have ampacities that are close to bare-conductor ampacities. The most significant difference is that covered conductors have less ability to withstand higher temperatures; the insulation degrades. Polyethylene is especially prone to damage, so it should not be operated above 75°C. EPR and XLPE may be operated up to 90°C.

Some utilities use two ratings, a "normal" ampacity with a 75°C design temperature and an "emergency" ampacity with a 90 or 100°C design. Conductors are selected for the normal rating, but operation is allowed to the emergency rating. Overhead circuits have considerable capability for overload, especially during cooler weather. We do not use relaying to trip "overloaded" circuits. At higher temperatures, conductors age more quickly but do not usually fail immediately.

2.7.1 Neutral Conductor Sizing

Because the neutral conductor carries less current than the phase conductors, utilities can use smaller neutral conductors. On three-phase circuits with

TABLE 2.16

Ampacities of All-Aluminum Conductor Covered with PE, XLPE, or EPR

AWG or kcmil	Stranding	Cover Thickness (mil)	Conductor Temp. = 75°C		Conductor Temp. = 90°C	
			25°C Ambient	40°C Ambient	25°C Ambient	40°C Ambient
6	7	30	105	85	120	105
4	7	30	140	110	160	135
2	7	45	185	145	210	180
1	7	45	210	170	245	210
1/0	7	60	240	195	280	240
2/0	7	60	280	225	325	280
3/0	7	60	320	255	375	320
4/0	7	60	370	295	430	370
4/0	19	60	375	295	430	370
266.8	19	60	430	340	500	430
336.4	19	60	500	395	580	495
397.5	19	80	545	430	635	545
477	37	80	615	480	715	610
556.5	37	80	675	530	785	675
636	61	95	725	570	850	725
795	61	95	835	650	980	835
1033.5	61	95	980	760	1150	985

Note: Emissivity = 0.91, absorptivity = 0.91; 30°N at 12 noon in clear atmosphere; wind speed = 2 ft/sec; elevation = sea level.

Source: Aluminum Association, *Ampacities for Aluminum and ACSR Overhead Electrical Conductors*, 1986.

balanced loading, the neutral carries almost no current. On single-phase circuits with a multigrounded neutral, the neutral normally carries 40 to 60% of the current (the earth carries the remainder).

On single-phase circuits, some utilities use fully rated neutrals, where the neutral and the phase are the same size. Some use reduced neutrals. The resistance of the neutral should be no more than twice the resistance of the phase conductor, and we are safer with a resistance less than 1.5 times the phase conductor, which is a conductivity or cross-sectional area of 2/3 the phase conductor. Common practice is to drop one to three gage sizes for the neutral: a 4/0 phase has a 2/0 neutral, or a 1/0 phase has a number 2 neutral. Dropping three gage sizes doubles the resistance, so we do not want to go any smaller than that.

On three-phase circuits, most utilities use reduced neutrals, dropping the area to about 25 to 70% of the phase conductor (and multiplying the resistance by 1.4 to 4).

Several other factors besides ampacity play a role in how small neutral conductors are:

- *Grounding* — A reduced neutral increases the overvoltages on the unfaulted phases during single line-to-ground faults (see Chapter 13). It also increases stray voltages.

- *Faults* — A reduced neutral reduces the fault current for single line-to-ground faults, which makes it more difficult to detect faults at far distances. Also, the reduced neutral is subjected to the same fault current as the phase, so impacts on burning down the neutral should be considered for smaller neutrals.

- *Secondary* — If the primary and secondary neutral are shared, the neutral must handle the primary and secondary unbalanced current (and have the mechanical strength to hold up the secondary phase conductors in triplex or quadraplex construction).

- *Mechanical* — On longer spans, the sag of the neutral should coordinate with the sag of the phases and the minimum ground clearances to ensure that spacing rules are not violated.

2.8 Secondaries

Utilities most commonly install *triplex* secondaries for overhead service to single-phase customers, where two insulated phase conductors are wrapped around the neutral. The neutral supports the weight of the conductors. Phase conductors are normally all-aluminum, and the neutral is all-aluminum, aluminum-alloy, or ACSR, depending on strength needs. Insulation is normally polyethylene, high-molecular weight polyethylene, or cross-linked polyethylene with thickness ranging from 30 to 80 mils (1.1 to 2 mm) rated for 600 V. Similarly for three-phase customers, quadraplex has three insulated phase conductors wrapped around a bare neutral. Table 2.17 shows characteristics of polyethylene triplex with an AAC neutral.

Triplex secondary ampacities depend on the temperature capability of the insulation. Polyethylene can operate up to 75°C. Cross-linked polyethylene and EPR can operate higher, up to 90°C. Table 2.18 shows ampacities for triplex when operated to each of these maximum temperatures. Quadraplex has ampacities that are 10 to 15% less than triplex of the same size conductor. Ampacities for open-wire secondary are the same as that for bare primary conductors.

Table 2.19 shows impedances of triplex. Two impedances are given: one for the 120-V loop and another for a 240-V loop. The 240-V loop impedance is the impedance to current flowing down one hot conductor and returning on the other. The 120-V loop impedance is the impedance to current down one hot conductor and returning in the neutral (and assuming no current returns through the earth). If the phase conductor and the neutral conductor are the same size, these impedances are the same. With a reduced neutral, the 120-V loop impedance is higher. Table 2.19 shows impedances for the reduced neutral size given; for a fully-rated neutral, use the 240-V impedance for the 120-V impedance.

TABLE 2.17

Typical Characteristics of Polyethylene-Covered AAC Triplex

Phase Conductor		ACSR Neutral Messenger		Reduced ACSR Neutral Messenger		AAC Neutral Messenger	
Size (Stranding)	Insulation Thickness, mil	Size (Stranding)	Rated Strength, lb	Size (Stranding)	Rated Strength, lb	Size (Stranding)	Rated Strength, lb
6 (1)	45	6 (6/1)	1190				
6 (7)	45	6 (6/1)	1190				
4 (1)	45	4 (6/1)	1860	6 (6/1)	1190	6 (7)	563
4 (7)	45	4 (6/1)	1860	6 (6/1)	1190	4 (7)	881
2 (7)	45	2 (6/1)	2850	4 (6/1)	1860	2 (7)	1350
1/0 (7)	60	1/0 (6/1)	4380	2 (6/1)	2853	1/0 (7)	1990
1/0 (19)	60	1/0 (6/1)	4380	2 (6/1)	2853	1/0 (7)	1990
2/0 (7)	60	2/0 (6/1)	5310	1 (6/1)	3550	2/0 (7)	2510
2/0 (19)	60	2/0 (6/1)	5310	1 (6/1)	3550		
3/0 (19)	60	3/0 (6/1)	6620	1/0 (6/1)	4380	3/0 (19)	3310
4/0 (19)	60	4/1 (6/1)	8350	2/0 (6/1)	5310	4/0 (19)	4020
336.4 (19)	80	336.4 (18/1)	8680	4/0 (6/1)	8350	336.4 (19)	6146

TABLE 2.18

Ampacities of All-Aluminum Triplex

Phase Conductor AWG	Strands	Conductor temp = 75°C		Conductor temp = 90°C	
		25°C Ambient	40°C Ambient	25°C Ambient	40°C Ambient
6	7	85	70	100	85
4	7	115	90	130	115
2	7	150	120	175	150
1/0	7	200	160	235	200
2/0	7	230	180	270	230
3/0	7	265	210	310	265
4/0	7	310	240	360	310

Note: Emissivity = 0.91, absorptivity = 0.91; 30°N at 12 noon in clear atmosphere; wind speed = 2 ft/sec; elevation = sea level.

Source: Aluminum Association, *Ampacities for Aluminum and ACSR Overhead Electrical Conductors*, 1986.

2.9 Fault Withstand Capability

When a distribution line short circuits, very large currents can flow for a short time until a fuse or breaker or other interrupter breaks the circuit. One important aspect of overcurrent protection is to ensure that the fault arc and fault currents do not cause further, possibly more permanent, damage. The two main considerations are:

TABLE 2.19

Typical Impedances of All-Aluminum Triplex Secondaries, Ω/1000 ft

Phase		Neutral		120-V Loop Impedance*		240-V Loop Impedance	
Size	Strands	Size	Strands	R_{S1}	X_{S1}	R_S	X_S
2	7	4	7	0.691	0.0652	0.534	0.0633
1	19	3	7	0.547	0.0659	0.424	0.0659
1/0	19	2	7	0.435	0.0628	0.335	0.0616
2/0	19	1	19	0.345	0.0629	0.266	0.0596
3/0	19	1/0	19	0.273	0.0604	0.211	0.0589
4/0	19	2/0	19	0.217	0.0588	0.167	0.0576
250	37	3/0	19	0.177	0.0583	0.142	0.0574
350	37	4/0	19	0.134	0.0570	0.102	0.0558
500	37	300	37	0.095	0.0547	0.072	0.0530

* With a full-sized neutral, the 120-V loop impedance is equal to the 240-V loop impedance.

Source: ABB Inc., *Distribution Transformer Guide*, 1995.

- *Conductor annealing* — From the substation to the fault location, all conductors in the fault-current path must withstand the heat generated by the short-circuit current. If the relaying or fuse does not clear the fault in time, the conductor anneals and loses strength.

- *Burndowns* — Right at the fault location, the hot fault arc can burn the conductor. If a circuit interrupter does not clear the fault in time, the arc will melt the conductor until it breaks apart.

For both annealing and arcing damage, we should design protection to clear faults before more damage is done. To do this, make sure that the time-current characteristics of the relay or fuse are faster than the time-current damage characteristics. Characteristics of annealing and arcing damage are included in the next two sections.

2.9.1 Conductor Annealing

During high currents from faults, conductors can withstand significant temperatures for a few seconds without losing strength. For all-aluminum conductors, assuming a maximum temperature of 340°C during faults is common. ACSR conductors can withstand even higher temperatures because short-duration high temperature does not affect the steel core. An upper limit of 645°C, the melting temperature of aluminum, is often assumed. For short-duration events, we ignore convection and radiation heat losses and assume that all heat stays in the conductor. With all heat staying in the conductor, the temperature is a function of the specific heat of the conductor material. Specific heat is the heat per unit mass required to raise the temperature by one degree Celsius (the specific heat of aluminum is 0.214 cal/g-°C). Considering the heat inputs and the conductor characteristics, the

TABLE 2.20

Conductor Thermal Data for Short-Circuit Limits

Conductor Material	λ, °C	K
Copper (97%)	234.0	0.0289
Aluminum (61.2%)	228.1	0.0126
6201 (52.5%)	228.1	0.0107
Steel	180.0	0.00327

Source: Southwire Company, *Overhead Conductor Manual*, 1994.

conductor temperature during a fault is related to the current (Southwire Company, 1994) as

$$\left(\frac{I}{1000A}\right)^2 t = K \log_{10}\left(\frac{T_2 + \lambda}{T_1 + \lambda}\right)$$

where

I = fault current, A
t = fault duration, sec
A = cross-sectional area of the conductor, kcmil
T_2 = conductor temperature from the fault, °C
T_1 = conductor temperature before the fault, °C
K = constant depending on the conductor, which includes the conductor's resistivity, density, and specific heat (see Table 2.20)
λ = inferred temperature of zero resistance, °C below zero (see Table 2.20)

If we set T_2 to the maximum allowable conductor temperature, we can find the maximum allowable I^2t characteristic for a given conductor. For all-aluminum conductors, with a maximum temperature, $T_2 = 340$°C, and an ambient of 40°C, the maximum allowable time-current characteristic for a given conductor size (Southwire Company, 1994) is

$$I^2t = (67.1A)^2$$

For ACSR with a maximum temperature of 640°C, the maximum allowable time-current characteristic for a given conductor size (Southwire Company, 1994) is

$$I^2t = (86.2A)^2$$

Covered conductors have more limited short-circuit capability because the insulation is damaged at lower temperatures. Thermoplastic insulations like polyethylene have a maximum short-duration temperature of 150°C. The thermoset insulations EPR and XLPE have a maximum short-duration tem-

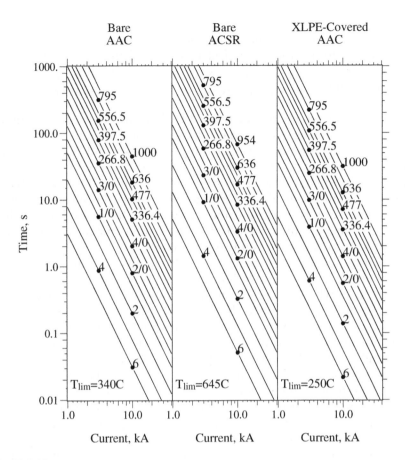

FIGURE 2.12
Annealing curves of bare AAC, ACSR, and covered AAC.

perature of 250°C. With these upper temperature limits (and $T_1 = 40°C$), the allowable time-current characteristics of aluminum conductors are:

$$\text{Polyethylene:} \quad I^2t = (43A)^2$$

$$\text{XLPE or EPR:} \quad I^2t = (56A)^2$$

Figure 2.12 compares short-circuit damage curves for various conductors.

2.9.2 Burndowns

Fault-current arcs can damage overhead conductors. The arc itself generates tremendous heat, and where an arc attaches to a conductor, it can weaken or burn conductor strands. On distribution circuits, two problem areas stand out:

1. *Covered conductor* — Covered conductor (also called tree wire or weatherproof wire) holds an arc stationary. Because the arc cannot move, burndowns happen faster than with bare conductors.
2. *Small bare wire on the mains* — Small bare wire (less than 2/0) is also susceptible to wire burndowns, especially if laterals are not fused.

Covered conductors are widely used to limit tree faults. Several utilities have had burndowns of covered conductor circuits when the instantaneous trip was not used or was improperly applied (Barker and Short, 1996; Short and Ammon, 1997). If a burndown on the main line occurs, all customers on the circuit will have a long interruption. In addition, it is a safety hazard. After the conductor breaks and falls to the ground, the substation breaker may reclose. After the reclosure, the conductor on the ground will probably not draw enough fault current to trip the station breaker again. This is a high-impedance fault that is difficult to detect.

A covered conductor is susceptible to burndowns because when a fault current arc develops, the covering prevents the arc from moving. The heat from the arc is what causes the damage. Although ionized air is a fairly good conductor, it is not as good as the conductor itself, so the arc gets very hot. On bare conductors, the arc is free to move, and the magnetic forces from the fault cause the arc to move (in the direction away from the substation; this is called *motoring*). The covering constricts the arc to one location, so the heating and melting is concentrated on one part of the conductor. If the covering is stripped at the insulators and a fault arcs across an insulator, the arc motors until it reaches the covering, stops, and burns the conductor apart at the junction. A party balloon, lightning, a tree branch, a squirrel — any of these can initiate the arc that burns the conductor down. Burndowns are most associated with lightning-caused faults, but it is the fault current arc, not the lightning, that burns most of the conductor.

Conductor damage is a function of the duration of the fault and the current magnitude. Burndown damage occurs much more quickly than conductor annealing that was analyzed in the previous section.

Although they are not as susceptible as covered conductors, bare conductors can also have burndowns. In tests of smaller bare conductors, Florida Power & Light Co. (FP&L) found that the hot gases from the arc anneal the conductor (Lasseter, 1956). They found surprisingly little burning from the arc; in fact, arcs could seriously degrade conductor strength even when there is no visible damage. Objects like insulators or tie wires absorb heat from the ionized gases and reduce the heat to the conductor.

What we would like to do is plot the arc damage characteristic as a function of time and current along with the time-current characteristics of the protective device (whether it be a fuse or a recloser or a breaker). Doing this, we can check that the protective device will clear the fault before the conductor is damaged. Figure 2.13 shows burndown damage characteristics for small bare ACSR conductors along with a 100 K lateral fuse element and a typical ground relay element. The fuse protects the conductors shown, but

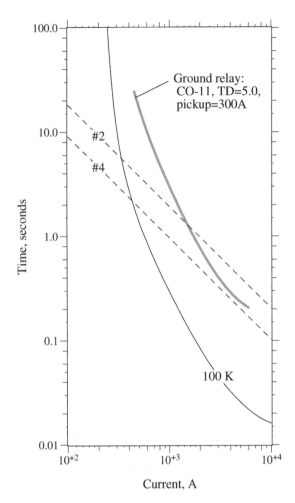

FIGURE 2.13
Bare-conductor ACSR threshold-of-damage curves along with the 100-K lateral fuse total clearing time and a ground relay characteristic. (Damage curves from [Lasseter, 1956].)

the ground relay does not provide adequate protection against damage for these conductors. These damage curves are based on FP&L's tests, where Lasseter reported that the threshold-of-damage was 25 to 50% of the average burndown time (see Table 2.21).

Such arc damage data for different conductor sizes as a function of time and current is limited. Table 2.22 summarizes burndown characteristics of some bare and covered conductors based on tests by Baltimore Gas & Electric (Goode and Gaertner, 1965). Figure 2.14 shows this same data on time-current plots along with a 100 K fuse total clearing characteristic. For conductor sizes not given, take the closest size given in Table 2.22, and scale the burndown time by the ratio of the given conductor area to the area of the desired conductor.

TABLE 2.21

The Burndown Characteristics of Several Small Bare
Conductors

Conductor	Threshold of Damage	Average Burndown Time
#4 AAAC	$t = \dfrac{4375}{I^{1.235}}$	$t = \dfrac{17500}{I^{1.235}}$
#4 ACSR	$t = \dfrac{800}{I^{0.973}}$	$t = \dfrac{3350}{I^{0.973}}$
#2 ACSR	$t = \dfrac{1600}{I^{0.973}}$	$t = \dfrac{3550}{I^{0.973}}$
#6 Cu	$t = \dfrac{410}{I^{0.909}}$	$t = \dfrac{1440}{I^{0.909}}$
#4 Cu	$t = \dfrac{500}{I^{0.909}}$	$t = \dfrac{1960}{I^{0.909}}$

Note: I = rms fault current, A; t = fault duration, sec.

Source: Lasseter, J.A., "Burndown Tests on Bare Conductor," *Electric Light and Power*, pp. 94–100, December 1956.

If covered conductor is used, consider the following options to limit burndowns:

- *Fuse saving* — Using a fuse blowing scheme can increase burndowns because the fault duration is much longer on the time-delay relay elements than on the instantaneous element. With fuse saving, the instantaneous relay element trips the circuit faster and reduces conductor damage.
- *Arc protective devices* (APDs) — These sacrificial masses of metal attach to the ends where the covering is stripped (Lee et al., 1980). The arc end attaches to the mass of metal, which has a large enough volume to withstand much more arcing than the conductor itself.
- *Fuse all taps* — Leaving smaller covered conductors unprotected is a sure way of burning down conductors.
- *Tighter fusing* — Not all fuses protect some of the conductor sizes used on taps. Faster fuses reduce the chance of burndowns.
- *Bigger conductors* — Bigger conductors take longer to burn down. Doubling the conductor cross-sectional area approximately doubles the time it takes to burn the conductor down.

Larger bare conductors are fairly immune to burndown. Smaller conductors used on taps are normally safe if protected by a fuse. The solutions for small bare conductors are

- *Fuse all taps* — This is the best option.

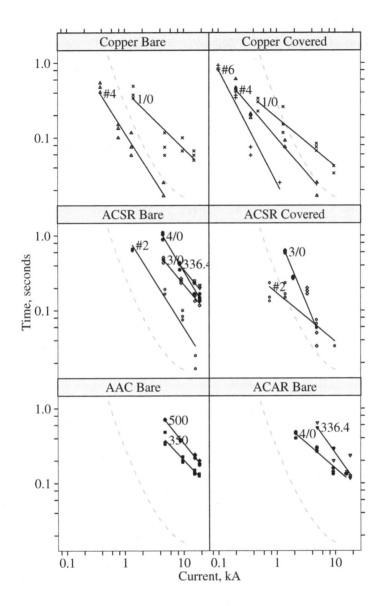

FIGURE 2.14
Burndown characteristics of various conductors. The dashed line is the total clearing time for a 100-K fuse. (Data from [Goode and Gaertner, 1965].)

- *Fuse saving* — The time-delay relay element may not protect smaller tap conductors. Faults cleared by an instantaneous element with fuse saving will not damage bare conductors. If fuse blowing is used, consider an alternative such as a high-set instantaneous or a delayed instantaneous (see Chapter 8 for more information).

TABLE 2.22

Burndown Characteristics of Various Conductors

		Duration, 60-Hz Cycles			
	Current, A	Min	Max	Other	Curvefit
#6 Cu covered	100	48.5	55.5	51	$t = 858/I^{1.51}$
	200	20.5	24.5	22	
	360	3.5	4.5	4.5	
	1140	1.5	1.5	1.5	
#4 Cu covered	200	26.5	36.5	28	$t = 56.4/I^{0.92}$
	360	11	12.5	12	
	1400	4.5	5.5	5.5	
	4900	1	1.5	1.5	
#4 Cu bare	380	24.5	32.5	28.5	$t = 641/I^{1.25}$
	780	6	9	8	
	1300	3.5	7	4.5	
	4600	1	1.5	1	
#2 ACSR covered	750	8	9	14	$t = 15.3/I^{0.65}$
	1400	10	9	14	
	4750	3.5	4.5	4	
	9800	2	2	NA	
#2 ACSR bare	1350	38	39	40	$t = 6718/I^{1.26}$
	4800	10	11.5	10	
	9600	4.5	5	6	
	15750	1	1.5	NA	
1/0 Cu covered	480	13.5	20	18	$t = 16.6/I^{0.65}$
	1300	7	15.5	9	
	4800	4	5	4.5	
	9600	2	2.5	2.5	
1/0 Cu bare	1400	20.5	29.5	22.5	$t = 91/I^{0.78}$
	4800	3.5	7	4.5	
	9600	4	6	6	
	15000	3	4	3.5	
3/0 ACSR covered	1400	35	38	37	$t = 642600/I^{1.92}$
	1900	16	17	16.5	
	3300	10	12	11	
	4800	2	3	3	
3/0 ACSR bare	4550	26	30.5	28.5	$t = 1460/I^{0.95}$
	9100	14	16	15	
	15500	8	9.5	8	
	18600	7	9	7	
4/0 ACAR bare	2100	24	29	28	$t = 80.3/I^{0.68}$
	4800	16	18	17.5	
	9200	8	9.5	8.5	
	15250	8	8	NA	
4/0 ACSR bare	4450	53	66	62	$t = 68810/I^{1.33}$
	8580	21	26	25	
	15250	10	14	NA	
	18700	8	10	8.5	
336.4-kcmil ACAR bare	4900	33	38.5	33	$t = 6610/I^{1.10}$
	9360	12	17.5	17	
	15800	8	8.5	8	
	18000	7	14	7.5	

TABLE 2.22 (Continued)

Burndown Characteristics of Various Conductors

		Duration, 60-Hz Cycles			
	Current, A	Min	Max	Other	Curvefit
336.4-kcmil ACSR bare	8425	25	26	26	$t = 2690/I^{0.97}$
	15200	10	15	14	
	18800	12	13	12	
350-kcmil AAC bare	4800	29	21	20	$t = 448/I^{0.84}$
	9600	11.5	13.5	12	
	15200	8	9	8.5	
	18200	8	7.5	7.5	
500-kcmil AAC bare	4800	42	43	42.5	$t = 2776/I^{0.98}$
	8800	22.5	23	22	
	15400	13	14.5	14	
	18400	11	12	10.5	

Source: Goode, W.B. and Gaertner, G.H., "Burndown Tests and Their Effect on Distribution Design," EEI T&D Meeting, Clearwater, FL, Oct. 14–15, 1965.

2.10 Other Overhead Issues

2.10.1 Connectors and Splices

Connectors and splices are often weak links in the overhead system, either due to hostile environment or bad designs or, most commonly, poor installation. Utilities have had problems with connectors, especially with higher loadings (Jondahl et al., 1991).

Most primary connectors use some sort of compression to join conductors (see Figure 2.15 for common connectors). Compression splices join two conductors together — two conductors are inserted in each end of the sleeve, and a compression tool is used to tighten the sleeve around the conductors. For conductors under tension, automatic splices are also available. Crews just insert the conductors in each end, and serrated clamps within the splice grip the conductor; with higher tension, the wedging action holds tighter.

For tapping a smaller conductor off of a larger conductor, many options are available. Hot-line clamps use a threaded bolt to hold the conductors together. Wedge connectors have a wedge driven between conductors held by a C-shaped body. Compression connectors (commonly called *squeezeons*) use dies and compression tools to squeeze together two conductors and the connector.

Good cleaning is essential to making a good contact between connector surfaces. Both copper and aluminum develop a hard oxide layer on the surface when exposed to air. While very beneficial in preventing corrosion, the oxide layer has high electrical resistance. Copper is relatively easy to brush clean. Aluminum is tougher; crews need to work at it harder, and a

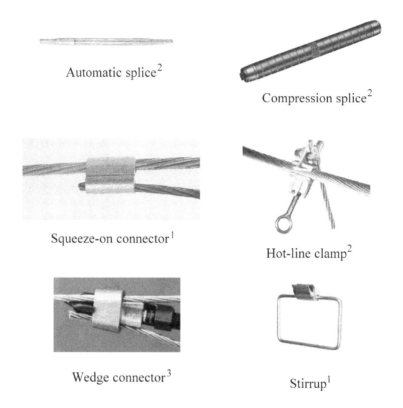

Automatic splice[2]

Compression splice[2]

Squeeze-on connector[1]

Hot-line clamp[2]

Wedge connector[3]

Stirrup[1]

FIGURE 2.15
Common distribution connectors. [1] Reprinted with the permission of Cooper Industries, Inc.
[2] Reprinted with the permission of Hubbell Power Systems, Inc. [3] Reprinted with the permission of Tyco Electronics Corporation.

shiny surface is no guarantee of a good contact. Aluminum oxidizes quickly, so crews should clean conductors just before attaching the connector. Without good cleaning, the temperatures developed at the hotspot can anneal the conductor, possibly leading to failure. Joint compounds are important; they inhibit oxidation and help maintain a good contact between joint surfaces.

Corrosion at interfaces can prematurely fail connectors and splices. Galvanic corrosion can occur quickly between dissimilar metals. For this reason, aluminum connectors are used to join aluminum conductors. Waterproof joint compounds protect conductors and joints from corrosion.

Aluminum expands and contracts with temperature, so swings in conductor temperature cause the conductor to creep with respect to the connector. This can loosen connectors and allow oxidation to develop between the connector and conductor. ANSI specifies a standard for connectors to withstand thermal cycling and mechanical stress (ANSI C119.4-1998).

Poor quality work leads to failures. Not using joint compound (or not using enough), inadequate conductor cleaning, misalignments, not fully inserting the conductor prior to compression, or using the wrong dies — any of these mistakes can cause a joint to fail prematurely.

Infrared thermography is the primary way utilities spot bad connectors. A bad connection with a high contact resistance operates at significantly higher temperatures than the surrounding conductor. While infrared inspections are easy for crews to do, they are not foolproof; they can miss bad connectors and falsely target good conductors. Infrared measurements are very sensitive to sunlight, line currents, and background colors. Temperature differences are most useful (but still not perfect indicators). Experience and visual checks of the connector can help identify false readings (such as glare due to sunlight reflection). A bad connector can become hot enough to melt the conductor, but often the conductor can resolidify, temporarily at a lower resistance. Infrared inspections can miss these bad connectors if they are in the resolidified stage. For compression splices, EPRI laboratory tests and field inspections found high success rates using hotstick-mounted resistance measuring devices that measure the resistance across a short section of the conductor (EPRI 1001913, 2001).

Short-circuit current can also damage inline connectors. Mechanical stresses and high currents can damage some splices and connectors. If an inline connector does not make solid contact at its interfaces to the conductor, hotspots can weaken and possibly break the connector or conductor. If the contact is poor enough to cause arcing, the arcing can quickly eat the connection away. Mechanical forces can also break an already weakened or corroded connector.

Hot-line clamps are popular connectors that crews can easily apply with hot-line tools. Threaded bolts provide compression between conductors. Hot-line clamps can become loose, especially if not installed correctly. Utilities have had problems with hot-line clamps attached directly to primary conductors, especially in series with the circuit (rather than tapped for a jumper to equipment) where they are subjected to the heat and mechanical forces of fault currents. Loose or high-resistance hot-line clamps can arc across the interface, quickly burning away the primary conductor.

Stirrups are widely used between the main conductor and a jumper to a transformer or capacitor bank. A stirrup is a bail or loop of wire attached to the main conductor with one or two compression connectors or hot-line connectors. Crews can quickly make a connection to the stirrup with hot-line clamps. The main reason for using the stirrup is to protect the main conductor from burndown. If tied directly to the main conductor, arcing across a poor connection can burn the main conductor down. If a poor hot-line clamp is connected to a stirrup, the stirrup may burn down, but the main line is protected. Also, any arcing when crews attach or detach the connector does not damage the main conductor, so stirrups are especially useful where jumpers may be put on and taken off several times. Using stirrups is reliable; a survey by the National Rural Electric Cooperative Association (NRECA) found that less than 10% of utilities have annual failure rates between 1 and 5%, and almost all of the remainder have failure rates less than 1% (RUS, 1996).

2.10.2 Radio Frequency Interference

Distribution line hardware can generate radio-frequency interference (RFI). Such interference can impact the AM and FM bands as well as VHF television broadcasts. Ham radio frequencies are also affected.

Most power-line noise is from arcs — arcs across gaps on the order of 1 mm, usually at poor contacts. These arcs can occur between many metallic junctions on power-line equipment. Consider two metal objects in close proximity but not quite touching. The capacitive voltage divider between the conducting parts determines the voltage differences between them. The voltage difference between two metallic pieces can cause an arc across a small gap. After arcing across, the gap can clear easily, and after the capacitive voltage builds back up, it can spark over again. These sparkovers radiate radio-frequency noise. Stronger radio-frequency interference is more likely from hardware closer to the primary conductors.

Arcing generates broadband radio-frequency noise from several kilohertz to over 1000 MHz. Above about 50 MHz, the magnitude of arcing RFI drops off. Power-line interference affects lower frequency broadcasts more than higher frequencies. The most common from low to high frequency are: AM radio (0.54 to 1.71 MHz), low-band VHF TV (channels 2 to 6, 54 to 88 MHz), FM radio (88.1 to 107.9 MHz), and high-band VHF TV (channels 7 to 13, 174 to 216 MHz). UHF (ultra-high frequencies, about 500 MHz) are only created right near the sparking source.

On an oscilloscope, arcing interference looks like a series of noise spikes clustered around the peaks of the sinusoidal power-frequency driving voltage (see Figure 2.16). Often power-line noise causes a raspy sound, usually with a 120-Hz characteristic. The "sound" of power-line noise varies depending on the length of the arcing gap, so interference cannot always be identified by a specific characteristic sound (Loftness, 1997).

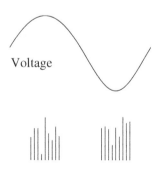

FIGURE 2.16
Arcing source creating radio-frequency interference.

Arcing across small gaps accounts for almost all radio-frequency interference created by utility equipment on distribution circuits. Arcing from corona can also cause interference, but distribution circuit voltages are too low for corona to cause significant interference. Radio interference is more common at higher distribution voltages.

Some common sources and solutions include [for more detail, see (Loftness, 1996; NRECA 90-30, 1992)]:

- *Loose or corroded hot-line clamps* — Replace the connector. After cleaning the conductor and applying fresh inhibitor replace the clamp with a new hot-line clamp or a wedge connector or a squeeze-on connector.

- *Loose nut and washer on a through bolt* — Commonly a problem on double-arming bolts between two crossarms; use lock washers and tighten.

- *Loose or broken insulator tie wire or incorrect tie wire* — Loose tie wires can cause arcing, and conducting ties on covered conductors generate interference; in either case, replace the tie wire.

- *Loose dead-end insulator units* — Replace, preferably with single-unit types. Semiconductive grease provides a short-term solution.

- *Loose metal staples on bonding or ground wires, especially near the top* — Replace with insulated staples (hammering in existing staples may only help for the short term).

- *Loose crossarm lag screw* — Replace with a larger lag screw or with a through bolt and lock washers.

- *Bonding conductors touching or nearly touching other metal hardware* — Separate by at least 1 in. (2.54 cm).

- *Broken or contaminated insulators* — Clean or replace.

- *Defective lightning arresters, especially gapped units* — Replace.

Most of these problems have a common characteristic: gaps between metals, often from loose hardware. Crews can fix most problems by tightening connections, separating metal hardware by at least 1 in., or bonding hardware together. Metal-to-wood interfaces are less likely to cause interference; a tree branch touching a conductor usually does not generate radio-frequency interference.

While interference is often associated with overhead circuits, underground lines can also generate interference. Again, look for loose connections, either primary or secondary such as in load-break elbows.

Interference from an arcing source can propagate in several ways: radiation, induction, and conduction. RFI can radiate from the arcing source just like a radio transmitter. It can conduct along a conductor and also couple inductively from one conductor to parallel conductors. Lower frequencies propagate farther; AM radio is affected over larger distances. Interference is roughly in inverse proportion to frequency at more than a few poles from the source.

Many different interference detectors are available; most are radios with directional antennas. Closer to the source, instruments can detect radio-frequency noise at higher and higher frequencies, so higher frequencies can help pinpoint sources. As you get closer to the source, follow the highest frequency that you can receive. (If you cannot detect interference at higher and higher frequencies as you drive along the line, you are probably going in the wrong direction.) Once a problem pole is identified, an ultrasonic detector with a parabolic dish can zero in on problem areas to identify where the arcing is coming from. Ultrasonic detectors measure ultra-high frequency sound waves (about 20 to 100 kHz) and give accurate direction to the source. Ultrasonic detectors almost require line-of-sight to the arcing source, so they do not help if the arcing is hidden. In such cases, the sparking may be internal to an enclosed device, or the RF could be conducted to the pole by a secondary conductor or riser pole. For even more precise location, crews can use hot-stick mounted detectors to identify exactly what's arcing.

Note that many other nonutility sources of radio-frequency interference exist. Many of these also involve intermittent arcing. Common power-frequency type sources include fans, light dimmers, fluorescent lights, loose wiring within the home or facility, and electrical tools such as drills. Other sources include defective antennas, amateur or CB radios, spark-plug ignitions from vehicles or lawn mowers, home computers, and garage door openers.

References

ABB, *Distribution Transformer Guide*, 1995.

Aluminum Association, *Ampacities for Aluminum and ACSR Overhead Electrical Conductors*, 1986.

Aluminum Association, *Aluminum Electrical Conductor Handbook*, 1989.

ANSI C119.4-1998, *Electric Connectors for Use Between Aluminum-to-Aluminum or Aluminum-to-Copper Bare Overhead Conductors*.

Barber, K., "Improvements in the Performance and Reliability of Covered Conductor Distribution Systems," International Covered Conductor Conference, Cheshire, U.K., January 1999.

Barker, P.P. and Short, T.A., "Findings of Recent Experiments Involving Natural and Triggered Lightning," IEEE/PES Transmission and Distribution Conference, Los Angeles, CA, 1996.

Black, W.Z. and Rehberg, R.L., "Simplified Model for Steady State and Real-Time Ampacity of Overhead Conductors," *IEEE Transactions on Power Apparatus and Systems*, vol. 104, pp. 29–42, October 1985.

Carson, J.R., "Wave Propagation in Overhead Wires with Ground Return," Bell System Technical Journal, vol. 5, pp. 539–54, 1926.

Clapp, A.L., *National Electrical Safety Code Handbook*, The Institute of Electrical and Electronics Engineers, Inc., 1997.

Clarke, E., *Circuit Analysis of AC Power Systems, II*, General Electric Company, 1950.

Ender, R.C., Auer, G.G., and Wylie, R.A., "Digital Calculation of Sequence Imped-ances and Fault Currents for Radial Primary Distribution Circuits," *AIEE Trans-actions on Power Apparatus and Systems*, vol. 79, pp. 1264–77, 1960.

EPRI 1001913, *Electrical, Mechanical, and Thermal Performance of Conductor Connections*, Electric Power Research Institute, Palo Alto, CA, 2001.

EPRI, *Transmission Line Reference Book: 345 kV and Above*, 2nd ed., Electric Power Research Institute, Palo Alto, CA, 1982.

Goode, W.B. and Gaertner, G.H., "Burndown Tests and Their Effect on Distribution Design," EEI T&D Meeting, Clearwater, FL, Oct. 14–15, 1965.

Harvey, J.R., "Effect of Elevated Temperature Operation on the Strength of Aluminum Conductors," IEEE/PES Winter Power Meeting Paper T 72 189–4, 1971. As cited by Southwire (1994).

House, H.E. and Tuttle, P.D., "Current Carrying Capacity of ACSR," *IEEE Transactions on Power Apparatus and Systems*, pp. 1169–78, February 1958.

IEEE C2-2000, *National Electrical Safety Code Handbook*, The Institute of Electrical and Electronics Engineers, Inc.

IEEE Std. 738-1993, *IEEE Standard for Calculating the Current-Temperature Relationship of Bare Overhead Conductors*.

Jondahl, D.W., Rockfield, L.M., and Cupp, G.M., "Connector Performance of New vs. Service Aged Conductor," IEEE/PES Transmission and Distribution Con-ference, 1991.

Kurtz, E.B., Shoemaker, T.M., and Mack, J.E., *The Lineman's and Cableman's Handbook*, McGraw Hill, New York, 1997.

Lasseter, J.A., "Burndown Tests on Bare Conductor," *Electric Light & Power*, pp. 94–100, December 1956.

Lat, M.V., "Determining Temporary Overvoltage Levels for Application of Metal Oxide Surge Arresters on Multigrounded Distribution Systems," *IEEE Trans-actions on Power Delivery*, vol. 5, no. 2, pp. 936–46, April 1990.

Lee, R.E., Fritz, D.E., Stiller, P.H., Kilar, L.A., and Shankle, D.F., "Prevention of Cov-ered Conductor Burndown on Distribution Circuits," American Power Confer-ence, 1980.

Loftness, M.O., *AC Power Interference Field Manual*, Percival Publishing, Tumwater, WA, 1996.

Loftness, M.O., "Power Line RF Interference — Sounds, Patterns, and Myths," *IEEE Transactions on Power Delivery*, vol. 12, no. 2, pp. 934–40, April 1997.

NRECA 90-30, *Power Line Interference: A Practical Handbook*, National Rural Electric Cooperative Association, 1992.

RUS 160-2, *Mechanical Design Manual for Overhead Distribution Lines*, United States Department of Agriculture, Rural Utilities Service, 1982.

RUS 1728F-803, *Specifications and Drawings for 24.9/14.4 kV Line Construction*, United States Department of Agriculture, Rural Utilities Service, 1998.

RUS, "Summary of Items of Engineering Interest," United States Department of Agriculture, Rural Utilities Service, 1996.

Short, T.A. and Ammon, R.A., "Instantaneous Trip Relay: Examining Its Role," *Trans-mission and Distribution World*, vol. 49, no. 2, 1997.

Smith, D.R., "System Considerations — Impedance and Fault Current Calculations," *IEEE Tutorial Course on Application and Coordination of Reclosers, Sectionalizers, and Fuses*, 1980. Publication 80 EHO157-8-PWR.

Southwire Company, *Overhead Conductor Manual*, 1994.

Stevenson, W.D., *Elements of Power System Analysis*, 2nd ed., McGraw Hill, New York, 1962.

Willis, H.L., *Power Distribution Planning Reference Book*, Marcel Dekker, New York, 1997.

Saying "You can't do that" to a Lineman is the same as saying "Hey, how about a contest?"

**Powerlineman law #23, By CD Thayer and other Power Linemen,
http://www.cdthayer.com/lineman.htm**

3

Underground Distribution

Much new distribution is underground. Underground distribution is much more hidden from view than overhead circuits, and is more reliable. Cables, connectors, and installation equipment have advanced considerably in the last quarter of the 20th century, making underground distribution installations faster and less expensive.

3.1 Applications

One of the main applications of underground circuits is for underground residential distribution (URD), underground branches or loops supplying residential neighborhoods. Utilities also use underground construction for substation exits and drops to padmounted transformers serving industrial or commercial customers. Other uses are crossings: river crossings, highway crossings, or transmission line crossings. All-underground construction — widely used for decades in cities — now appears in more places.

Underground construction is expensive, and costs vary widely. Table 3.1 shows extracts from one survey of costs done by the CEA; the two utilities highlighted differ by a factor of ten. The main factors that influence underground costs are:

- *Degree of development* — Roads, driveways, sidewalks, and water pipes — these and other obstacles slow construction and increase costs.

- *Soil condition* — Rocks and frozen ground increase overtime pay for cable crews.

- *Urban, suburban, or rural* — Urban construction is more difficult not only because of concrete, but also because of traffic. Rural construction is generally the least expensive per length, but lengths are long.

- *Conduit* — Concrete-encased ducts cost more than direct-buried conduits, which cost more than preassembled flexible conduit, which cost more than directly buried cable with no conduits.

TABLE 3.1

Comparison of Costs of Different Underground Constructions at Different Utilities

Utility	Construction	$/ft[a]
TAU	Rural or urban, 1 phase, #2 Al, 25 kV, trenched, direct buried	6.7
	Rural, 3 phase, #2 Al, 25 kV, trenched, direct buried	13.4
	Urban commercial, 3 phase, #2 Al, 25 kV, trenched, direct buried	13.4
	Urban express, 3 phase, 500-kcmil Al, 25 kV, trenched, direct buried	23.5
WH	Urban, 1 phase, 1/0 Al, 12.5 kV, trenched, conduit	84.1
	Urban commercial, 3 phase, 1/0 Al, 12.5 kV, trenched, conduit	117.7
	Urban express, 3 phase, 500-kcmil Cu, 12.5 kV, trenched, conduit	277.4

[a] Converted assuming that one 1991 Canadian dollar equals 1.1 U.S. dollars in 2000.

Source: CEA 274 D 723, *Underground Versus Overhead Distribution Systems,* Canadian Electrical Association, 1992.

- *Cable size and materials* — The actual cable cost is a relatively small part of many underground applications. A 1/0 aluminum full-neutral 220-mil TR-XLPE cable costs just under $2 per ft; with a 500-kcmil conductor and a one-third neutral, the cable costs just under $4 per ft.

- *Installation equipment* — Bigger machines and machines more appropriate for the surface and soil conditions ease installations.

3.1.1 Underground Residential Distribution (URD)

A classic underground residential distribution circuit is an underground circuit in a loop arrangement fed at each end from an overhead circuit (see Figure 3.1). The loop arrangement allows utilities to restore customers more quickly; after crews find the faulted section, they can reconfigure the loop and isolate any failed section of cable. This returns power to all customers. Crews can delay replacing or fixing the cable until a more convenient time or when suitable equipment arrives. Not all URD is configured in a loop. Utilities sometimes use purely radial circuits or circuits with radial taps or branches.

Padmounted transformers step voltage down for delivery to customers and provide a sectionalizing point. The elbow connectors on the cables (pistol grips) attach to bushings on the transformer to maintain a dead-front — no exposed, energized conductors. To open a section of cable, crews can simply pull an elbow off of the transformer bushing and place it on a parking stand, which is an elbow bushing meant for holding an energized elbow connector.

Elbows and other terminations are available with continuous-current ratings of 200 or 600 A (IEEE Std. 386-1995). Load-break elbows are designed to break load; these are only available in 200-A ratings. Without load-break capability, crews should of course only disconnect the elbow if the cable is deenergized. Elbows normally have a test point where crews can check if

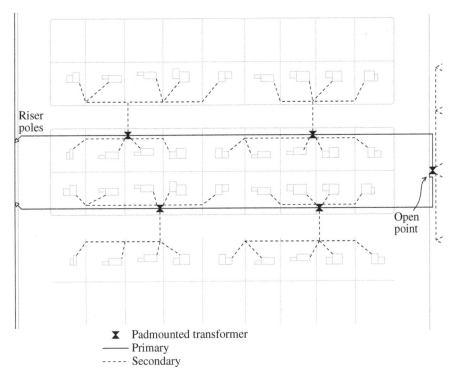

X Padmounted transformer
——— Primary
- - - - - Secondary

FIGURE 3.1
An example front-lot underground residential distribution (URD) system.

the cable is live. Elbows are also tested to withstand ten cycles of fault current, with 200-A elbows tested at 10 kA and 600-A elbows tested at 25 kA (IEEE Std. 386-1995).

The interface between the overhead circuit and the URD circuit is the riser pole. At the riser pole (or a dip pole or simply a dip), cable terminations provide the interface between the insulated cable and the bare overhead conductors. These pothead terminations grade the insulation to prevent excessive electrical stress on the insulation. Potheads also keep water from entering the cable, which is critical for cable reliability. Also at the riser pole are expulsion fuses, normally in cutouts. Areas with high short-circuit current may also have current-limiting fuses. To keep lightning surges from damaging the cable, the riser pole should have arresters right across the pothead with as little lead length as possible.

Underground designs for residential developments expanded dramatically in the 1970s. Political pressure coupled with technology improvements were the driving forces behind underground distribution. The main developments — direct-buried cables and padmounted transformers having load-break elbows — dramatically reduced the cost of underground distribution to close to that of overhead construction. In addition to improving the visual landscape, underground construction improves reliability. Underground residential distribution has had difficulties, especially high cable failure rates. In the late 1960s and early 1970s, given the durability of plastics, the poly-ethylene cables installed at that time were thought to have a life of at least 50 years. In practice, cables failed at a much higher rate than expected, enough so that many utilities had to replace large amounts of this cable.

According to Boucher (1991), 72% of utilities use front-lot designs for URD. With easier access and fewer trees and brush to clear, crews can more easily install cables along streets in the front of yards. Customers prefer rear-lot service, which hides padmounted transformers from view. Back-lot place-ment can ease siting issues and may be more economical if lots share rear property lines. But with rear-lot design, utility crews have more difficulty accessing cables and transformers for fault location, sectionalizing, and repair.

Of those utilities surveyed by Boucher (1991), 85% charge for underground residential service, ranging from $200 to $1200 per lot (1991 dollars). Some utilities charge by length, which ranges from $5.80 to $35.00 per ft.

3.1.2 Main Feeders

Whether urban, suburban, or even rural, all parts of a distribution circuit can be underground, including the main feeder. For reliability, utilities often configure an underground main feeder as a looped system with one or more tie points to other sources. Switching cabinets or junction boxes serve as tie points for tapping off lateral taps or branches to customers. These can be in handholes, padmounted enclosures, or pedestals above ground. Three-phase circuits can also be arranged much like URD with sections of cable run between three-phase padmounted transformers. As with URD, the pad-mounted transformers serve as switching stations.

Although short, many feeders have an important underground section — the substation exit. Underground substation exits make substations eas-ier to design and improve the aesthetics of the substation. Because they are at the substation, the source of a radial circuit, substation exits are critical for reliability. In addition, the loading on the circuit is higher at the sub-station exit than anywhere else; the substation exit may limit the entire circuit's ampacity. Substation exits are not the place to cut corners. Some strategies to reduce the risks of failures or to speed recovery are: concrete-enclosed ducts to help protect cables, spare cables, overrated cables, and good surge protection.

While not as critical as substation exits, utilities use similar three-phase underground dips to cross large highways or rivers or other obstacles. These are designed in much the same way as substation exits.

3.1.3 Urban Systems

Underground distribution has reliably supplied urban systems since the early 1900s. Cables are normally installed in concrete-encased duct banks beneath streets, sidewalks, or alleys. A duct bank is a group of parallel ducts, usually with four to nine ducts but often many more. Ducts may be precast concrete sections or PVC encased in concrete. Duct banks carry both primary and secondary cables. Manholes every few hundred feet provide access to cables. Transformers are in vaults or in the basements of large buildings.

Paper-insulated lead-covered (PILC) cables dominated urban applications until the late 20th century. Although a few utilities still install PILC, most use extruded cable for underground applications. In urban applications, copper is more widely used than in suburban applications. Whether feeding secondary networks or other distribution configurations, urban circuits may be subjected to heavy loads.

"Vertical" distribution systems are necessary in very tall buildings. Medium-voltage cable strung up many floors feed transformers within a building. Submarine cables are good for this application since their protective armor wire provides support when a cable is suspended for hundreds of feet.

3.1.4 Overhead vs. Underground

Overhead or underground? The debate continues. Both designs have advantages (see Table 3.2). The major advantage of overhead circuits is cost; an underground circuit typically costs anywhere from 1 to 2.5 times the equivalent overhead circuit (see Table 3.3). But the cost differences vary wildly, and it's often difficult to define "equivalent" systems in terms of performance. Under the right conditions, some estimates of cost report that cable installations can be less expensive than overhead lines. If the soil is easy to dig, if the soil has few rocks, if the ground has no other obstacles like water pipes or telephone wires, then crews may be able to plow in cable faster and for less cost than an overhead circuit. In urban areas, underground is almost the only choice; too many circuits are needed, and above-ground space is too expensive or just not available. But urban duct-bank construction is expensive on a per-length basis (fortunately, circuits are short in urban appli-

TABLE 3.2

Overhead vs. Underground: Advantages of Each

Overhead	Underground
Cost — Overhead's number one advantage. Significantly less cost, especially initial cost.	*Aesthetics* — Underground's number one advantage. Much less visual clutter.
Longer life — 30 to 50 years vs. 20 to 40 for new underground works.	*Safety* — Less chance for public contact.
Reliability — Shorter outage durations because of faster fault finding and faster repair.	*Reliability* — Significantly fewer short and long-duration interruptions.
Loading — Overhead circuits can more readily withstand overloads.	*O&M* — Notably lower maintenance costs (no tree trimming).
	Longer reach — Less voltage drop because reactance is lower.

TABLE 3.3

Comparison of Underground Construction Costs with Overhead Costs

Utility	Construction		$/ft[a]	Underground to overhead ratio
Single-Phase Lateral Comparisons				
NP	Overhead	1/0 AA, 12.5 kV, phase and neutral	8.4	
NP	Underground	1/0 AA, 12.5 kV, trenched, in conduit	10.9	1.3
APL	Overhead	Urban, #4 ACSR, 14.4 kV	2.8	
APL	Underground	Urban, #1 AA, 14.4 kV, trenched, direct buried	6.6	2.4
Three-Phase Mainline Comparisons				
NP	Overhead	Rural, 4/0 AA, 12.5 kV	10.3	
NP	Underground	Rural, 1/0 AA, 12.5 kV, trenched, in conduit	17.8	1.7
NP	Overhead	Urban, 4/0 AA, 12.5 kV	10.9	
NP	Underground	Urban, 4/0 AA, 12.5 kV, trenched, in conduit	17.8	1.6
APL	Overhead	Urban, 25 kV, 1/0 ACSR	8.5	
APL	Underground	Urban, 25 kV, #1 AA, trenched, direct buried	18.8	2.2
EP	Overhead	Urban, 336 ACSR, 13.8 kV	8.7	
EP	Underground	Urban residential, 350 AA, 13.8 kV, trenched, direct buried	53.2	6.1
EP	Underground	Urban commercial, 350 AA, 13.8 kV, trenched, direct buried	66.8	7.6

[a] Converted assuming that one 1991 Canadian dollar equals 1.1 U.S. dollars in 2000.

Source: CEA 274 D 723, *Underground Versus Overhead Distribution Systems*, Canadian Electrical Association, 1992.

cations). On many rural applications, the cost of underground circuits is difficult to justify, especially on long, lightly loaded circuits, given the small number of customers that these circuits feed.

Aesthetics is the main driver towards underground circuits. Especially in residential areas, parks, wildlife areas, and scenic areas, visual impact is important. Undergrounding removes a significant amount of visual clutter. Overhead circuits are ugly. It is possible to make overhead circuits less ugly with tidy construction practices, fiberglass poles instead of wood, keeping poles straight, tight conductor configurations, joint use of poles to reduce the number of poles, and so on. Even the best though, are still ugly, and many older circuits look awful (weathered poles tipped at odd angles, crooked crossarms, rusted transformer tanks, etc.).

Underground circuits get rid of all that mess, with no visual impacts in the air. Trees replace wires, and trees don't have to be trimmed. At ground level, instead of poles every 150 ft (many having one or more guy wires) urban construction has no obstacles, and URD-style construction has just padmounted transformers spaced much less frequently. Of course, for maximum benefit, all utilities must be underground. There is little improvement to undergrounding electric circuits if phone and cable television are still strung on poles (i.e., if the telephone wires are overhead, you might as well have the electric lines there, too).

While underground circuits are certainly more appealing when finished, during installation construction is messier than overhead installation. Lawns, gardens, sidewalks, and driveways are dug up; construction lasts longer; and the installation "wounds" take time to heal. These factors don't matter much when installing circuits into land that is being developed, but it can be upsetting to customers in an existing, settled community.

Underground circuits are more reliable. Overhead circuits typically fault about 90 times/100 mi/year; underground circuits fail less than 10 times/100 mi/year. Because overhead circuits have more faults, they cause more voltage sags, more momentary interruptions, and more long-duration interruptions. Even accounting for the fact that most overhead faults are temporary, overhead circuits have more permanent faults that lead to long-duration circuit interruptions. The one disadvantage of underground circuits is that when they do fail, finding the failure is harder, and fixing the damage or replacing the equipment takes longer. This can partially be avoided by using loops capable of serving customers from two directions, by using conduits for faster replacement, and by using better fault location techniques. Underground circuits are much less prone to the elements. A major hurricane may drain an overhead utility's resources, crews are completely tied up, customer outages become very long, and cleanup costs are a major cost to utilities. However, underground circuits are not totally immune from the elements. In "heat storms," underground circuits are prone to rashes of failures. Underground circuits have less overload capability than overhead circuits; failures increase with operating temperature.

In addition to less storm cleanup, underground circuits require less periodic maintenance. Underground circuits don't require tree trimming, easily the largest fraction of most distribution operations and maintenance budgets. The CEA (1992) estimated that underground system maintenance averaged 2% of system plant investment whereas overhead systems averaged 3 to 4%, or as much as twice that of underground systems.

Underground circuits are safer to the public than overhead circuits. Overhead circuits are more exposed to the public. Kites, ladders, downed wires, truck booms — despite the best public awareness campaigns, these still expose the public to electrocution from overhead lines. Don't misunderstand; underground circuits still have dangers, but they're much less than on overhead circuits. For the public, dig-ins are the most likely source of contact. For utility crews, both overhead and underground circuits offer dangers that proper work practices must address to minimize risks.

We cannot assume that underground infrastructure will last as long as overhead circuits. Early URD systems failed at a much higher rate than expected. While most experts believe that modern underground equipment is more reliable, it is still prudent to believe that an overhead circuit will last 40 years, while an underground circuit will only last 30 years.

Overhead vs. underground is not an all or nothing proposition. Many systems are hybrids; some schemes are:

- *Overhead mainline with underground taps* — The larger, high-current conductors are overhead. If the mains are routed along major roads, they have less visual impact. Lateral taps down side roads and into residential areas, parks, and shopping areas are underground. Larger primary equipment like regulators, reclosers, capacitor banks, and automated switches are installed where they are more economical — on the overhead mains. Because the mainline is a major contributor to reliability, this system is still less reliable than an all-underground system.

- *Overhead primary with underground secondary* — Underground secondary eliminates some of the clutter associated with overhead construction. Eliminating much of the street and yard crossings keeps the clutter to the pole-line corridor. Costs are reasonable because the primary-level equipment is still all overhead.

Converting from overhead to underground is costly, yet there are locations and situations where it is appropriate for utilities and their customers. Circuit extensions, circuit enhancements to carry more load, and road-rebuilding projects — all are opportunities for utilities and communities to upgrade to underground service.

3.2   Cables

At the center of a cable is the phase conductor, then comes a semiconducting conductor shield, the insulation, a semiconducting insulation shield, the neutral or shield, and finally a covering jacket. Most distribution cables are single conductor. Two main types of cable are available: concentric-neutral cable and power cable. Concentric-neutral cable normally has an aluminum conductor, an extruded insulation, and a concentric neutral (Figure 3.2 shows a typical construction). A concentric neutral is made from several copper wires wound concentrically around the insulation; the concentric neutral is a true neutral, meaning it can carry return current on a grounded system. Underground residential distribution normally has concentric-neutral cables; concentric-neutral cables are also used for three-phase mainline applications and three-phase power delivery to commercial and industrial customers. Because of their widespread use in URD, concentric-neutral cables are often called URD cables. Power cable has a copper or aluminum phase

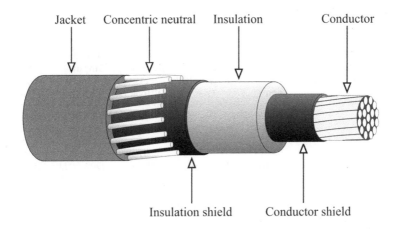

Jacket Concentric neutral Insulation Conductor

Insulation shield Conductor shield

FIGURE 3.2
A concentric neutral cable, typically used for underground residential power delivery.

conductor, an extruded insulation, and normally a thin copper tape shield. On utility distribution circuits, power cables are typically used for mainline feeder applications, network feeders, and other high current, three-phase applications. Many other types of medium-voltage cable are available. These are sometimes appropriate for distribution circuit application: three-conductor power cables, armored cables, aerial cables, fire-resistant cables, extra flexible cables, and submarine cables.

3.2.1 Cable Insulation

A cable's insulation holds back the electrons; the insulation allows cables with a small overall diameter to support a conductor at significant voltage. A 0.175-in. (4.5-mm) thick polymer cable is designed to support just over 8 kV continuously; that's an average stress of just under 50 kV per in. (20 kV/cm). In addition to handling significant voltage stress, insulation must withstand high temperatures during heavy loading and during short circuits and must be flexible enough to work with. For much of the 20th century, paper insulation dominated underground application, particularly PILC cables. The last 30 years of the 20th century saw the rise of polymer-insulated cables, polyethylene-based insulations starting with high-molecular weight polyethylene (HMWPE), then cross-linked polyethylene (XLPE), then tree-retardant XLPE and also ethylene-propylene rubber (EPR) compounds.

Table 3.4 compares properties of TR-XLPE, EPR, and other insulation materials. Some of the key properties of cable insulation are:

- *Dielectric constant* (ε, also called permittivity) — This determines the cable's capacitance: the dielectric constant is the ratio of the capacitance with the insulation material to the capacitance of the same

TABLE 3.4

Properties of Cable Insulations

	Dielectric Constant 20°C	Loss Angle Tan δ at 20°C	Volume Resistivity Ω-m	Annual Dielectric Loss[a] W/1000 ft	Unaged Impulse Strength V/mil	Water Absorption ppm
PILC	3.6	0.003	10^{11}	N/A	1000–2000	25
PE	2.3	0.0002	10^{14}	N/A		100
XLPE	2.3	0.0003	10^{14}	8	3300	350
TR-XLPE	2.4	0.001	10^{14}	10	3000	<300
EPR	2.7–3.3	0.005–0.008	10^{13}–10^{14}	28–599	1200–2000	1150–3200

[a] For a typical 1/0 15-kV cable.

Copyright © 2001. Electric Power Research Institute. 1001894. *EPRI Power Cable Materials Selection Guide*. Reprinted with permission.

configuration in free space. Cables with higher capacitance draw more charging current.

- *Volume resistivity* — Current leakage through the insulation is a function of the insulation's dc resistivity. Resistivity decreases as temperature increases. Modern insulation has such high resistivity that very little resistive current passes from the conductor through the insulation.

- *Dielectric losses* — Like a capacitor, a cable has dielectric losses. These losses are due to dipole movements within the polymer or by the movement of charge carriers within the insulation. Dielectric losses contribute to a cable's resistive leakage current. Dielectric losses increase with frequency and temperature and with operating voltage.

- *Dissipation factor* (also referred to as the loss angle, loss tangent, tan δ, and approximate power factor) — The dissipation factor is the ratio of the resistive current drawn by the cable to the capacitive current drawn (I_R/I_X). Because the leakage current is normally low, the dissipation factor is approximately the same as the power factor:

$$\text{pf} = I_R / |I| = I_R / \sqrt{I_R^2 + I_X^2} \approx I_R / I_X = \text{dissipation factor}$$

Paper-Insulated Lead-Covered (PILC) Cables. Paper-insulated cables have provided reliable underground power delivery for decades. Paper-insulated lead-sheathed cable has been the dominant cable configuration, used mainly in urban areas. PILC cables have kraft-paper tapes wound around the conductor that are dried and impregnated with insulating oil. A lead sheath is one of the best moisture blocks: it keeps the oil in and keeps water out. Paper cables are normally rated to 85°C with an emergency rating up to 105°C (EPRI TR-105502, 1995). PILC cables have held up astonishingly well; many 50-year-old cables are still in service with almost new insulation capability. While PILC has had very good reliability, some utilities are concerned about its present day failure, not because of bad design or application, but because the in-service stock is so old. Moisture ingress, loss of oil, and thermal stresses — these are the three main causes of PILC failure (EPRI 1000741, 2000). Water decreases the dielectric strength (especially when the cable is hot) and increases the dielectric losses (further heating the cable). Heat degrades the insulating capability of the paper, and if oil is lost, the paper's insulating capability declines. PILC use has declined but still not disappeared. Some utilities continue to use it, especially to supply urban networks. Utilities use less PILC because of its high cost, work difficulties, and environmental concerns. Splicing also requires significant skill, and working with the lead sheath requires environmental and health precautions.

Polyethylene (PE). Most modern cables have polymer insulation extruded around the conductor — either polyethylene derivatives or ethylene-propylene properties. Polyethylene is a tough, inexpensive polymer with good electrical properties. Most distribution cables made since 1970 are based on some variation of polyethylene. Polyethylene is an ethylene polymer, a long string or chain of connected molecules. In polyethylene, some of the polymer chains align in crystalline regions, which give strength and moisture resistance to the material. Other regions have nonaligned polymer chains — these amorphous regions give the material flexibility but are permeable to gas and moisture and are where impurities locate. Polyethylene is a thermoplastic. When heated and softened, the polymer chains break apart (becoming completely amorphous); as it cools, the crystalline regions reform, and the material returns to its original state. Polyethylene naturally has high density and excellent electrical properties with a volume resistivity of greater than 10^{14} Ω-m and an impulse insulation strength of over 2700 V/mil.

High-Molecular Weight Polyethylene (HMWPE). High-molecular weight polyethylene is polyethylene that is stiffer, stronger, and more resistant to chemical attack than standard polyethylene. Insulations with higher molecular weights (longer polymer chains) generally have better electrical properties. As with standard polyethylene, HMWPE insulation is a thermoplastic rated to 75°C. Polyethylene softens considerably as temperature increases. Since plastics are stable and seem to last forever, when utilities first installed HMWPE in the late 1960s and early 1970s, utilities and manufacturers expected long life for polyethylene cables. In practice, failure rates increased dramatically after as little as 5 years of service. The electrical insulating strength (the dielectric strength) of HMWPE was degraded by water treeing, an electrochemical degradation driven by the presence of water and voltage. Polyethylene also degrades quickly under partial discharges; once partial discharges start, they can quickly eat away the insulation. Because of high failure rates, HMWPE insulation is off the market now, but utilities still have many miles of this cable in the ground.

Cross-Linked Polyethylene (XLPE). Cross-linking agents are added that form bonds between polymer chains. The cross-linking bonds interconnect the chains and make XLPE semi-crystalline and add stiffness. XLPE is a thermoset: the material is vulcanized (also called "cured"), irreversibly creating the cross-linking that sets when the insulation cools. XLPE has about the same insulation strengths as polyethylene, is more rigid, and resists water treeing better than polyethylene. Although not as bad as HMWPE, pre-1980s XLPE has proven susceptible to premature failures because of water treeing. XLPE has higher temperature ratings than HWMPE; cables are rated to 90°C under normal conditions and 130°C for emergency conditions.

Tree-Retardant Cross-Linked Polyethylene (TR-XLPE). This has adders to XLPE that slow the growth of water trees. Tree-retardant versions of XLPE have almost totally displaced XLPE in medium-voltage cables. Various compounds when added to XLPE reduce its tendency to grow water trees under voltage. These additives tend to slightly reduce XLPE's electrical properties, slightly increase dielectric losses, and slightly lower initial insulation strength (but much better insulation strength when aged). While there is no standard industry definition of TR-XLPE, different manufacturers offer XLPE compounds with various adders that reduce tree growth. The oldest and most widely used formulation was developed by Union Carbide (now Dow); their HFDA 4202 tree-retardant XLPE maintains its insulation strength better in accelerated aging tests (EPRI TR-108405-V1, 1997) and in field service (Katz and Walker, 1998) than standard XLPE.

Ethylene-Propylene Rubber (EPR). EPR compounds are polymers made from ethylene and propylene. Manufacturers offer different ethylene-propylene formulations, which collectively are referred to as EPR. EPR compounds are thermoset, normally with a high-temperature steam curing process that sets cross-linking agents. EPR compounds have high concentrations of clay fillers that provide its stiffness. EPR is very flexible and rubbery. When new, EPR only has half of the insulation strength as XLPE, but as it ages, its insulation strength does not decrease nearly as much as that of XLPE. EPR is naturally quite resistant to water trees, and EPR has a proven reliable record in the field. EPR has very good high-temperature performance. Although soft, it deforms less at high temperature than XLPE and maintains its insulation strength well at high temperature (Brown, 1983). Most new EPR cables are rated to 105°C under normal conditions and to 140°C for emergency conditions, the MV-105 designation per UL Standard 1072. (Historically, both XLPE and EPR cables were rated to 90°C normal and 130°C emergency.) In addition to its use as cable insulation, most splices and joints are made of EPR compounds. EPR has higher dielectric losses than XLPE; depending on the particular formulation, EPR can have two to three times the losses of XLPE to over ten times the losses of XLPE. These losses increase the cost of operation over its lifetime. While not as common or as widely used as XLPE in the utility market, EPR dominates for medium-voltage industrial applications.

TR-XLPE vs. EPR: which to use? Of the largest investor-owned utilities 56% specify TR-XLPE cables, 24% specify EPR, and the remainder specify a mix (Dudas and Cochran, 1999). Trends are similar at rural cooperatives. In a survey of the co-ops with the largest installed base of underground cable, 42% specify TR-XLPE, 34% specify EPR, and the rest specify both (Dudas and Rodgers, 1999). When utilities specify both EPR and TR-XLPE, commonly EPR is used for 600-A three-phase circuits, and TR-XLPE is used for 200-A applications like URD. Each cable type has advocates. TR-XLPE is less expensive and has lower losses. EPR's main feature is its long history of reliability and water-tree resistance. EPR is also softer (easier to handle) and has a higher temperature rating (higher ampacity). Boggs and Xu (2001) show how EPR and TR-XLPE are becoming more similar: EPR compounds are being designed that have fewer losses; tree-retardant additives to XLPE make the cable more tree resistant at the expense of increasing its water absorption and slightly increasing losses.

Cables have a voltage rating based on the line-to-line voltage. Standard voltage ratings are 5, 8, 15, 25, and 35 kV. A single-phase circuit with a nominal voltage of 7.2 kV from line to ground must use a 15-kV cable, not an 8-kV cable (because the line-to-line voltage is 12.47 kV).

Within each voltage rating, more than one insulation thickness is available. Standards specify three levels of cable insulation based on how the cables are applied. The main factor is grounding and ability to clear line-to-ground faults in order to limit the overvoltage on the unfaulted phases. The standard levels are (AEIC CS5-94, 1994):

- *100 percent level* — Allowed where line-to-ground faults can be cleared quickly (at least within one minute); normally appropriate for grounded circuits
- *133 percent level* — Where line-to-ground faults can be cleared within one hour; normally can be used on ungrounded circuits

Standards also define a 173% level for situations where faults cannot be cleared within one hour, but manufacturers typically offer the 100 and 133% levels as standard cables; higher insulation needs can be met by a custom order or going to a higher voltage rating. Table 3.5 shows standard insulation thicknesses for XLPE and EPR for each voltage level. In addition to protecting against temporary overvoltages, thicker insulations provide higher insulation to lightning and other overvoltages and reduce the chance of failure from water tree growth. For 15-kV class cables, Boucher (1991) reported that 59% of utilities surveyed in North America use 100% insulation (175-mil).

TABLE 3.5

Usual Insulation Thicknesses for XLPE or EPR Cables Based on Voltage and Insulation Level

Voltage Rating, kV	Insulation Thickness, Mil (1 mil = 0.001 in. = 0.00254 cm)	
	100% Level	133% Level
8	115	140
15	175	220
25	260	320
35	345	420

At 25 and 35 kV, the surveyed utilities more universally use 100% insulation (88 and 99%, respectively). Dudas and Cochran (1999) report similar trends in a survey of practices of the 45 largest investor-owned utilities: at 15 kV, 69% of utilities specified 100% insulation; at 25 and 35 kV, over 99% of utilities specified 100% insulation.

3.2.2 Conductors

For underground residential distribution (URD) applications, utilities normally use aluminum conductors; Boucher (1991) reported that 80% of utilities use aluminum (alloy 1350); the remainder, copper (annealed, soft). Copper is more prevalent in urban duct construction and in industrial applications. Copper has lower resistivity and higher ampacity for a given size; aluminum is less expensive and lighter. Cables are often stranded to increase their flexibility (solid conductor cables are available for less than 2/0). ASTM class B stranding is the standard stranding. Class C has more strands for applications requiring more flexibility. Each layer of strands is wound in an opposite direction. Table 3.6 shows diameters of available conductors.

3.2.3 Neutral or Shield

A cable's shield, the metallic barrier that surrounds the cable insulation, holds the outside of the cable at (or near) ground potential. It also provides a path for return current and for fault current. The shield also protects the cable from lightning strikes and from current from other fault sources. The metallic shield is also called the *sheath*.

A concentric neutral — a shield capable of carrying unbalanced current — has copper wires wound helically around the insulation shield. The concentric neutral is expected to carry much of the unbalanced load current, with the earth carrying the rest. For single-phase cables, utilities normally use a "full neutral," meaning that the resistance of the neutral equals that of the phase conductor. Also common is a "one-third neutral," which has a resistance that is three times that of the phase conductor. In a survey of underground distribution practices, Boucher (1991) reported that full neutrals dominated for residential application, and reduced neutrals are used more for commercial and feeder applications (see Figure 3.3).

Power cables commonly have 5-mil thick copper tape shields. These are wrapped helically around the cable with some overlap. In a tape-shield cable, the shield is not normally expected to carry unbalanced load current. As we will see, there is an advantage to having a higher resistance shield: the cable ampacity can be higher because there is less circulating current. Shields are also available that are helically wound wires (like a concentric neutral but with smaller wires).

Whether wires or tapes, cable shields and neutrals are copper. Aluminum corrodes too quickly to perform well in this function. Early unjacketed cables

TABLE 3.6

Conductor Diameters

Size	Solid Diameter, in.	Class B stranding Strands	Class B stranding Diameter, in.
24	0.0201	7	0.023
22	0.0253	7	0.029
20	0.032	7	0.036
19	0.035	7	0.041
18	0.0403	7	0.046
16	0.0508	7	0.058
14	0.0641	7	0.073
12	0.0808	7	0.092
10	0.1019	7	0.116
9	0.1144	7	0.13
8	0.1285	7	0.146
7	0.1443	7	0.164
6	0.162	7	0.184
5	0.1819	7	0.206
4	0.2043	7	0.232
3	0.2294	7	0.26
2	0.2576	7	0.292
1	0.2893	19	0.332
1/0	0.3249	19	0.373
2/0	0.3648	19	0.419
3/0	0.4096	19	0.47
4/0	0.46	19	0.528
250		37	0.575
300		37	0.63
350		37	0.681
400		37	0.728
500		37	0.813
600		61	0.893
750		61	0.998
1000		61	1.152
1250		91	1.289
1500		91	1.412
1750		127	1.526
2000		127	1.632
2500		127	1.824

normally had a coating of lead-tin alloy to prevent corrosion. Cable neutrals still corroded. Dudas (1994) reports that in 1993, 84% of utilities specified a bare copper neutral rather than a coated neutral.

The longitudinally corrugated (LC) shield improves performance for fault currents and slows down water entry. The folds of a corrugated copper tape are overlapped over the cable core. The overlapping design allows movement and shifting while also slowing down water entry. The design performs better for faults because it is thicker than a tape shield, so it has less resistance, and it tends to distribute current throughout the shield rather than keeping it in a few strands.

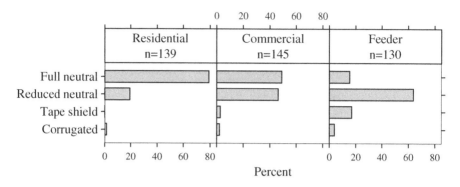

FIGURE 3.3
Surveyed utility use of cable neutral configurations for residential, commercial, and feeder applications. (Data from [Boucher, 1991].)

3.2.4 Semiconducting Shields

In this application, semiconducting means "somewhat conducting": the material has some resistance (limited to a volume resistivity of 500 Ω-m [ANSI/ICEA S-94-649-2000, 2000; ANSI/ICEA S-97-682-2000, 2000]), more than the conductor and less than the insulation. Semiconducting does not refer to nonlinear resistive materials like silicon or metal oxide; the resistance is fixed; it does not vary with voltage. Also called screens or semicons, these semiconducting shields are normally less than 80 mil. The resistive material evens out the electric field at the interface between the phase conductor and the insulation and between the insulation and the neutral or shield. Without the shields, the electric field gradient would concentrate at the closest inter-faces between a wire and the insulation; the increased localized stress could break down the insulation. The shields are made by adding carbon to a normally insulating polymer like EPR or polyethylene or cross-linked poly-ethylene. The conductor shield is normally about 20 to 40 mil thick; the insulation shield is normally about 40 to 80 mil thick. Thicker shields are used on larger diameter cables.

Semiconducting shields are important for smoothing out the electric field, but they also play a critical role in the formation of water trees. The most dangerous water trees are vented trees, those that start at the interface between the insulation and the semiconducting shield. Treeing starts at voids and impurities at this boundary. "Supersmooth" shield formulations have been developed to reduce vented trees (Burns, 1990). These mixtures use finer carbon particles to smooth out the interface. Under accelerated aging tests, cables with supersmooth semiconducting shields outperformed cables with standard semiconducting shields.

Modern manufacturing techniques can extrude the semiconducting con-ductor shield, the insulation, and the semiconducting insulation shield in one pass. Using this *triple* extrusion provides cleaner, smoother contact between layers than extruding each layer in a separate pass.

A note on terminology: a *shield* is the conductive layer surrounding another part of the cable. The conductor shield surrounds the conductor; the insulation shield surrounds the insulation. Used generically, shield refers to the metallic shield (the sheath). Commonly, the metallic shield is called the neutral, the shield, or the sheath. Sometimes, the sheath is used to mean the outer part of the cable, whether conducting or not conducting.

3.2.5 Jacket

Almost all new cables are jacketed, and the most common jacket is an encapsulating jacket (it is extruded between and over the neutral wires). The jacket provides some (but not complete) protection against water entry. It also provides mechanical protection for the neutral. Common LLDPE jackets are 50 to 80 mil thick.

Bare cable, used frequently in the 1970s, had a relatively high failure rate (Dedman and Bowles, 1990). Neutral corrosion was often cited as the main reason for the higher failure rate. At sections with a corroded neutral, the ground return current can heat spots missing neutral strands. Dielectric failure, not neutral corrosion, is still the dominant failure mode (Gurniak, 1996). Without the jacket, water enters easily and accelerates water treeing, which leads to premature dielectric failure.

Several materials are used for jackets. Polyvinyl chloride (PVC) was one of the earliest jacketing materials and is still common. The most common jacket material is made from linear low-density polyethylene (LLDPE). PVC has good jacketing properties, but LLDPE is even better in most regards: mechanical properties, temperature limits, and water entry. Moisture passes through PVC jacketing more than ten times faster than it passes through LLDPE. LLDPE starts to melt at 100°C; PVC is usually more limited, depending on composition. Low-density polyethylene resists abrasion better and also has a lower coefficient of friction, which makes it easier to pull through conduit.

Semiconducting jackets are also available. Semiconducting jackets provide the grounding advantages of unjacketed cable, while also blocking moisture and physically protecting the cable. When direct buried, an exposed neutral provides an excellent grounding conductor. The neutral in contact with the soil helps improve equipment grounding and improves protection against surges. A semiconducting jacket has a resistivity equivalent to most soils (less than 100 Ω-m), so it transfers current to the ground the same as an unjacketed cable. NRECA (1993) recommends not using a semiconducting jacket for two reasons. First, semiconducting jackets let more water pass through than LLDPE jackets. Second, the semiconducting jacket could contribute to corrosion. The carbon in the jacket (which makes the jacket semiconducting) is galvanic to the neutral and other nearby metals; especially with water in the cable, the carbon accelerates neutral corrosion. Other nearby objects in the ground such as ground rods or pipes can also corrode more rapidly from the carbon in the jacket.

3.3   Installations and Configurations

Just as there are many different soil types and underground applications, utilities have many ways to install underground cable. Some common installation methods include [see NRECA RER Project 90-8 (1993) for more details]:

- *Trenching* — This is the most common way to install cables, either direct-buried or cables in conduit. After a trench is dug, cable is installed, backfill is added and tamped, and the surface is restored. A trenching machine with different cutting chains is available for use on different soils. Backhoes also help with trenching.

- *Plowing* — A cable plow blade breaks up and lifts the earth as it feeds a cable into the furrow. Plowing eliminates backfilling and disturbs the surface less than trenching. NRECA reports that plowing is 30 to 50% less expensive than trenching (NRECA RER Project 90-8, 1993). Plowed cables may have lower ampacity because of air pockets between the cable and the loose soil around the cable. Heat cannot transfer as effectively from the cable to the surrounding earth.

- *Boring* — A number of tunneling technologies are available to drill under roads or even over much longer distances with guided, fluid-assisted drill heads.

Utilities also have a number of installation options, each with tradeoffs:

- *Direct buried* — Cables are buried directly in the earth. This is the fastest and least expensive installation option. Its major disadvantage is that cable replacement or repair is difficult.

- *Conduit* — Using conduit allows for quicker replacement or repair. Rigid PVC conduit is the most common conduit material; steel and HDPE and fiberglass are also used. Cables in conduit have less ampacity than direct-buried cables.

- *Direct buried with a spare conduit* — Burying a cable with a spare conduit provides provisions for repair or upgrades. Crews can pull another cable through the spare conduit to increase capacity or, if the cable fails, run a replacement cable through the spare conduit and abandon the failed cable. Normally, when the cable is plowed in, the conduit is coilable polyethylene.

- *Concrete-encased conduit* — Most often used in urban construction, conduit is encased in concrete. Concrete protects the conduit, resisting collapse due to shifting earth. The concrete also helps prevent dig-ins.

- *Preassembled cable in conduit* — Cable with flexible conduit can be purchased on reels, which crews can plow into the ground together.

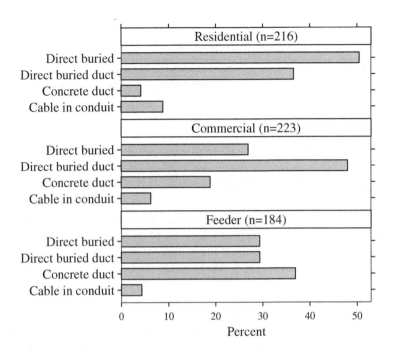

FIGURE 3.4
Surveyed utility cable installation configurations for residential, commercial, and feeder applications. (Data from [Boucher, 1991].)

The flexible conduit is likely to be more difficult to pull cable through, especially if the conduit is not straight. Flexible conduit is also not as strong as rigid conduit; the conduit can collapse due to rocks or other external forces.

Utilities are split between using direct-buried cable and conduits or ducts for underground residential applications. Conduits are used more for three-phase circuits, for commercial service, and for main feeder applications (see Figure 3.4). Conduit use is rising as shown by a more recent survey in Table 3.7. In a survey of the rural cooperatives with the most underground distribution, Dudas and Rodgers (1999) reported that 80% directly bury cable.

With conduits, customers have less outage time because cables can be replaced or repaired more quickly. In addition, replacement causes much less trouble for customers. Replacement doesn't disturb driveways, streets, or lawns; crews can concentrate their work at padmounted gear, rather than spread out along entire cable runs; and crews are less likely to tie up traffic. Conduit costs more than direct buried cable initially, typically from 25 to 50% more for PVC conduit (but this ranges widely depending on soil conditions and obstacles in or on the ground). Cable in flexible conduit may be slightly less than cable in rigid conduit. While directly buried cable has lower initial costs, lifetime costs can be higher than conduit depending on economic

TABLE 3.7

Surveyed Utility Use of Cable Duct Installations

	Percent of Cable Miles with Each Configuration	
	1998 Installed	Planned for the Future
Direct buried	64.6	46.9
Installed in conduit sections	25.5	37.9
Preassembled cable in conduit	7.1	11.7
Direct buried with a spare conduit	1.1	0.5
Continuous lengths of PE tubing	1.7	3.0

Source: Tyner, J. T., "Getting to the Bottom of UG Practices," *Transmission & Distribution World*, vol. 50, no. 7, pp. 44–56, July 1998.

assumptions and assumptions on how long cables will last or if they will need to be upgraded. Some utilities use a combination approach; most cable is direct buried, but ducts are used for road crossings and other obstacles.

The National Electrical Safety Code requires that direct-buried cable have at least 30 in. (0.75 m) of cover (IEEE C2-1997). Typically, trench depths are at least 36 in.

If communication cables are buried with primary power cables, extra rules apply. For direct-buried cable with an insulating jacket, the NESC requires that the neutral must have at least one half of the conductivity of the phase conductor (IEEE C2-1997) (it must be a one-half neutral or a full neutral).

Some urban applications are constrained by small ducts: 3, 3.5, or 4-in. diameters. These ducts were designed to hold three-conductor paper-insulated lead-sheathed cables which have conductors squashed in a sector shape for a more compact arrangement. Insulation cannot be extruded over these shapes, so obtaining an equivalent replacement cable with extruded insulation is difficult. Manufacturers offer thinner cables to meet these applications. For triplex cable, the equivalent outside diameter is 2.155 times the diameter of an individual cable. So, to fit in a 3-in. duct, an individual cable must be less than 1.16 in. in diameter to leave a 1/2-in. space (see Table 3.8 for other duct sizes). Some cable offered as "thin-wall" cable has slightly reduced insulation. For 15-kV cable, the smallest insulation thicknesses range between 150 and 165 mil as compared to the standard 175 mil (EPRI 1001734, 2002) (the ICEA allows 100% 15-kV cable insulation to range from 165 to 205

TABLE 3.8

Maximum Cable Diameters for Small Conduits Using PILC or Triplexed Cables that Leave 1/2-in. Pulling Room

Duct Size, in.	Largest Three-Conductor 15-kV PILC	Maximum Cable Diameter for Triplex Construction, in.	Largest Standard Construction Triplexed 15-kV Copper Cable
3.0	350 kcmil	1.16	3/0
3.5	750 kcmil	1.39	350 kcmil
4.0	1000 kcmil	1.62	500 kcmil

mil (ANSI/ICEA S-97-682-2000, 2000)). One manufacturer has proposed reduced insulation thicknesses based on the fact that larger conductors have lower peak voltage stress on the insulation than smaller conductors (Cinquemani et al., 1997), for example, 110-mil insulation at 15 kV for 4/0 through 750 kcmil. The maximum electric field (EPRI 1001734, 2002) is given by

$$E_{max} = \frac{2V}{d\ln(D/d)}$$

where

E_{max} = maximum electric field, V/mil (or other distance unit)
V = operating or rated voltage to neutral, V
d = inside diameter of the insulation, mil (or other distance unit)
D = outside diameter of the insulation in the same units as d

So, a 750-kcmil cable with 140-mil insulation has about the same maximum voltage stress as a 1/0 cable with 175-mil insulation at the same voltage. Nevertheless, most manufacturers are reluctant to trim the primary insulation too much, fearing premature failure due to water treeing. In addition to slightly reduced insulation, thin-wall cables are normally compressed copper and have thinner jackets and thinner semiconducting shields around the conductor and insulation. EPRI has also investigated other polymers for use in thin-wall cables (EPRI TR-111888, 2000). Their investigations found promising results with novel polymer blends that could achieve insulation strengths that are 30 to 40% higher than XLPE. These tests suggest promise, but more work must be done to improve the extrusion of these materials.

3.4 Impedances

3.4.1 Resistance

Cable conductor resistance is an important part of impedance that is used for fault studies and load flow studies. Resistance also greatly impacts a cable's ampacity. The major variable that affects resistance is the conductor's temperature; resistance rises with temperature. Magnetic fields from alternating currents also reduce a conductor's resistance relative to its dc resistance. At power frequencies, skin effect is only apparent for large conductors and proximity effect only occurs for conductors in very tight configurations. The starting point for resistance calculations is the dc resistance. From there, we can adjust for temperature and for frequency effects. Table 3.9 shows the dc resistances of several common conductors used for cables.

Resistance increases with temperature as

TABLE 3.9

dc Resistance at 25°C in Ω/1000 ft

Size	Aluminum Solid	Aluminum Class-B Stranded	Uncoated Copper Solid	Uncoated Copper Class-B Stranded	Coated Copper Solid	Coated Copper Class-B Stranded
24			26.2		27.3	
22			16.5		17.2	
20			10.3	10.5	10.7	11.2
19			8.21		8.53	
18			6.51	6.64	6.77	7.05
16			4.1	4.18	4.26	4.44
14	4.22		2.57	2.62	2.68	2.73
12	2.66	2.7	1.62	1.65	1.68	1.72
10	1.67	1.7	1.02	1.04	1.06	1.08
9	1.32	1.35	0.808	0.824	0.831	0.857
8	1.05	1.07	0.641	0.654	0.659	0.679
7	0.833	0.85	0.508	0.518	0.523	0.539
6	0.661	0.674	0.403	0.41	0.415	0.427
5	0.524	0.535	0.319	0.326	0.329	0.339
4	0.415	0.424	0.253	0.259	0.261	0.269
3	0.33	0.336	0.201	0.205	0.207	0.213
2	0.261	0.267	0.159	0.162	0.164	0.169
1	0.207	0.211	0.126	0.129	0.13	0.134
1/0	0.164	0.168	0.1	0.102	0.103	0.106
2/0	0.13	0.133	0.0795	0.0811	0.0814	0.0843
3/0	0.103	0.105	0.063	0.0642	0.0645	0.0668
4/0	0.082	0.0836	0.05	0.0509	0.0512	0.0525
250		0.0708		0.0431		0.0449
300		0.059		0.036		0.0374
350		0.0505		0.0308		0.032
400		0.0442		0.027		0.0278
500		0.0354		0.0216		0.0222
600		0.0295		0.018		0.0187
750		0.0236		0.0144		0.0148
1000		0.0177		0.0108		0.0111
1250		0.0142		0.00863		0.00888
1500		0.0118		0.00719		0.0074
1750		0.0101		0.00616		0.00634
2000		0.00885		0.00539		0.00555
2500		0.00715		0.00436		0.00448

Note: × 5.28 for Ω/mi or × 3.28 for Ω/km.

$$R_{t2} = R_{t1} \frac{M + t_2}{M + t_1}$$

where

R_{t2} = resistance at temperature t_2 given, °C

R_{t1} = resistance at temperature t_1 given, °C

M = a temperature coefficient for the given material

= 228.1 for aluminum

= 234.5 for soft-drawn copper

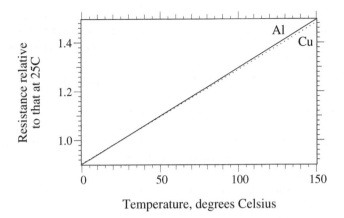

FIGURE 3.5
Resistance change with temperature.

Both copper and aluminum change resistivity at about the same rate as shown in Figure 3.5.

The ac resistance of a conductor is the dc resistance increased by a skin effect factor and a proximity effect factor

$$R = R_{dc}(1 + Y_{cs} + Y_{cp})$$

where
R_{dc} = dc resistance at the desired operating temperature, $\Omega/1000$ ft
Y_{cs} = skin-effect factor
Y_{cp} = proximity effect factor

The skin-effect factor is a complex function involving Bessel function solutions. The following polynomial approximates the skin-effect factor (Anders, 1998):

$$Y_{cs} = \frac{x_s^4}{192 + 0.8x_s^4} \qquad \text{for } x_s \leq 2.8$$

$$Y_{cs} = -0.136 - 0.0177x_s + 0.0563x_s^2 \qquad \text{for } 2.8 < x_s \leq 3.8$$

$$Y_{cs} = \frac{x_s}{2\sqrt{2}} - \frac{11}{15} \qquad \text{for } 3.8 < x_s$$

where

$$x_s = 0.02768\sqrt{\frac{f \cdot k_s}{R_{dc}}}$$

f = frequency, Hz

k_s = skin effect constant = 1 for typical conductors in extruded cables, may be less than one for paper cables that are dried and impregnated and especially those with round segmental conductors [see Neher and McGrath (1957) or IEC (1982)].

R_{dc} = dc resistance at the desired operating temperature, $\Omega/1000$ ft

For virtually all applications at power frequency, x_s is < 2.8.

With a conductor in close proximity to another current-carrying conductor, the magnetic fields from the adjacent conductor force current to flow in the portions of the conductor most distant from the adjacent conductor (with both conductors carrying current in the same direction). This magnetic field effect increases the effective ac resistance. The proximity effect factor is approximately (Anders, 1998; IEC 287, 1982):

$$Y_{cp} = ay^2\left(0.312y^2 + \frac{1.18}{a+0.27}\right)$$

where

$$a = \frac{x_p^4}{192+0.8x_p^4}, \quad y = \frac{d_c}{s}$$

$$x_p = 0.02768\sqrt{\frac{f \cdot k_p}{R_{dc}}}$$

d_c = conductor diameter

s = distance between conductor centers

k_p = proximity effect constant = 1 for typical conductors in extruded cables; may be < 1 for paper cables that are dried and impregnated and especially those with round segmental conductors [see Neher and McGrath (1957) or IEC (1982)].

At power frequencies, we can ignore proximity effect if the spacing exceeds ten times the conductor diameter (the effect is less than 1%).

Table 3.10 and Table 3.11 show characteristics of common cable conductors.

3.4.2 Impedance Formulas

Smith and Barger (1972) showed that we can treat a multi-wire concentric neutral as a uniform sheath; further work by Lewis and Allen (1978) and by Lewis, Allen, and Wang (1978) simplified the calculation of the representation of the concentric neutral. Following the procedure and nomenclature of Smith (1980) and Lewis and Allen (1978), we can find a cable's sequence impedances from the self and mutual impedances of the cable phase and neutral conductors as

TABLE 3.10

Characteristics of Aluminum Cable Conductors

Conductor	Stranding	GMR, in.	ac/dc Resistance Ratio	Resistances, Ω/1000 ft		
				dc at 25°C	ac at 25°C	ac at 90°C
2	7	0.105	1	0.2660	0.2660	0.3328
1	19	0.124	1	0.2110	0.2110	0.2640
1/0	19	0.139	1	0.1680	0.1680	0.2102
2/0	19	0.156	1	0.1330	0.1330	0.1664
3/0	19	0.175	1	0.1050	0.1050	0.1314
4/0	19	0.197	1	0.0836	0.0836	0.1046
250	37	0.216	1.01	0.0707	0.0714	0.0893
350	37	0.256	1.01	0.0505	0.0510	0.0638
500	37	0.305	1.02	0.0354	0.0361	0.0452
750	61	0.377	1.05	0.0236	0.0248	0.0310
1000	61	0.435	1.09	0.0177	0.0193	0.0241

TABLE 3.11

Characteristics of Copper Cable Conductors

Conductor	Stranding	GMR, in.	ac/dc Resistance Ratio	Resistances, Ω/1000 ft		
				dc at 25°C	ac at 25°C	ac at 90°C
2	7	0.105	1	0.1620	0.1620	0.2027
1	19	0.124	1	0.1290	0.1290	0.1614
1/0	19	0.139	1	0.1020	0.1020	0.1276
2/0	19	0.156	1.01	0.0810	0.0818	0.1023
3/0	19	0.175	1.01	0.0642	0.0648	0.0811
4/0	19	0.197	1.01	0.0510	0.0515	0.0644
250	37	0.216	1.01	0.0431	0.0435	0.0545
350	37	0.256	1.03	0.0308	0.0317	0.0397
500	37	0.305	1.06	0.0216	0.0229	0.0286
750	61	0.377	1.13	0.0144	0.0163	0.0204
1000	61	0.435	1.22	0.0108	0.0132	0.0165

$$Z_{11} = Z_{aa} - Z_{ab} - \frac{(Z_{ax} - Z_{ab})^2}{Z_{xx} - Z_{ab}}$$

$$Z_{00} = Z_{aa} + 2Z_{ab} - \frac{(Z_{ax} + 2Z_{ab})^2}{Z_{xx} + 2Z_{ab}}$$

The self and mutual impedances in the sequence equations are found with

$$Z_{aa} = R_\phi + R_e + jk_1 \log_{10} \frac{D_e}{GMR_\phi}$$

$$Z_{ab} = R_e + jk_1 \log_{10} \frac{D_e}{GMD_\phi}$$

$$Z_{xx} = R_N + R_e + jk_1 \log_{10} \frac{D_e}{GMR_N}$$

$$Z_{ax} = R_e + jk_1 \log_{10} \frac{D_e}{DN2}$$

where the self and mutual impedances with earth return are:

Z_{aa} = self impedance of each phase conductor

Z_{ab} = the mutual impedance between two conductors (between two phases, between two neutral, or between a phase and a neutral)

Z_{ax} = the mutual impedance between a phase conductor and its concentric neutral (or sheath)

Z_{xx} = self impedance of each concentric neutral (or shield)

and

R_ϕ = resistance of the phase conductor, Ω/distance

R_N = resistance of the neutral (or shield), Ω/distance

k_1 = 0.2794f/60 for outputs in Ω/mi

  = 0.0529f/60 for outputs in Ω/1000 ft

f = frequency, Hz

GMR_ϕ = geometric mean radius of the phase conductor, in. (see Table 3.12)

GMD_ϕ = geometric mean distance between the phase conductors, in.

  $= \sqrt[3]{d_{AB}d_{BC}d_{CA}}$

  = 1.26 d_{AB} for a three-phase line with flat configuration, either horizontal or vertical, when $d_{AB} = d_{BC} = 0.5d_{CA}$

  = the cable's outside diameter for triplex cables

  = 1.15 times the cable's outside diameter for cables cradled in a duct

d_{ij} = distance between the center of conductor i and the center of conductor j, in. (see Figure 3.6)

R_e = resistance of the earth return path

  = 0.0954(f/60)Ω/mi

  = 0.01807(f/60)Ω/1000 ft

D_e = $25920\sqrt{\rho/f}$ = equivalent depth of the earth return current, in.

ρ = earth resistivity, Ω-m

GMR_N = geometric mean radius of the sheath or neutral. For single-conductor cables with tape or lead sheaths, set GMR_N equal to the average radius of the sheath. For cables with a multi-wire concentric neutral, use $GMR_N = \sqrt[n]{0.7788nDN2^{(n-1)}r_n}$ where n is the number of neutrals and r_n is the radius of each neutral, in.

TABLE 3.12

Geometric Mean Radius (GMR) of Class B Stranded
Copper and Aluminum Conductors

Size	Stranding	GMR, in.		
		Round	Compressed	Compact
8	7	0.053		
6	7	0.067		
4	7	0.084		
2	7	0.106	0.105	
1	19	0.126	0.124	0.117
1/0	19	0.141	0.139	0.131
2/0	19	0.159	0.156	0.146
3/0	19	0.178	0.175	0.165
4/0	19	0.200	0.197	0.185
250	37	0.221	0.216	0.203
350	37	0.261	0.256	0.240
500	37	0.312	0.305	0.287
750	61	0.383	0.377	0.353
1000	61	0.442	0.435	0.413

Source: Southwire Company, *Power Cable Manual*, 2nd ed., 1997.

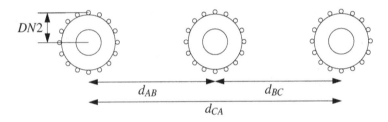

FIGURE 3.6
Cable dimensions for calculating impedances.

$DN2$ = effective radius of the neutral = the distance from the center of the
phase conductor to the center of a neutral strand, in.

Smith (1980) reported that assuming equal GMR_N and $DN2$ for cables from
1/0 to 1000 kcmil with one-third neutrals is accurate to 1%.

For single-phase circuits, the zero and positive-sequence impedances are
the same:

$$Z_{11} = Z_{00} = Z_{aa} - \frac{Z_{ax}^2}{Z_{xx}}$$

This is the loop impedance, the impedance to current flow through the phase
conductor that returns in the neutral and earth. The impedances of two-
phase circuits are more difficult to calculate (see Smith, 1980).

The sheath resistances depend on whether it is a concentric neutral, a tape shield, or some other configuration. For a concentric neutral, the resistance is approximately (ignoring the lay of the neutral):

$$R_{neutral} = \frac{R_{strand}}{n}$$

where
   R_{strand} = resistance of one strand, in Ω/unit distance
       n = number of strands

A tape shield's resistance (Southwire Company, 1997) is

$$R_{shield} = \frac{\rho_c}{A_s}$$

where
   ρ_c = resistivity of the tape shield, Ω-cmil/ft = 10.575 for uncoated copper
       at 25°C
   A_s = effective area of the shield in circular mil

$A_s = 4b \cdot d_m \cdot \sqrt{\dfrac{50}{100 - L}}$

   b = thickness of the tape, mil
   d_m = mean diameter outside of the metallic shield, mil
   L = lap of the tape shield in percent (normally 10 to 25%)

Normally, we can use dc resistance as the ac resistance for tape shields or concentric neutrals. The skin effect is very small because the shield conductors are thin (skin effect just impacts larger conductors). We should adjust the sheath resistance for temperature; for copper conductors, the adjustment is:

$$R_{t2} = R_{t1} \frac{234.5 + t_2}{234.5 + t_1}$$

where
   R_{t2} = resistance at temperature t_2 given in °C
   R_{t1} = resistance at temperature t_1 given in °C

These calculations are simplifications. More advanced models, normally requiring a computer, can accurately find each element in the full impedance matrix. For most load-flow calculations, this accuracy is not needed, though access to user-friendly computer models allows quicker results than calculating the equations shown here. For evaluating switching transients and some ampacity problems or configurations with several cables, we sometimes need more sophisticated models [see Amateni (1980) or Dommel (1986) for analytical details].

In a cable, the neutral tightly couples with the phase. Phase current induces neutral voltages that force circulating current in the neutrals. With balanced, positive-sequence current in the three phases and with symmetrical conductors, the neutral current (Lewis and Allen, 1978; Smith and Barger, 1972) is

$$I_{X1} = -\frac{Z_{ax} - Z_{ab}}{Z_{xx} - Z_{ab}} I_a$$

which is

$$I_{X1} = -\frac{j0.0529 \log_{10} \dfrac{d_{ab}}{DN2}}{R_N + j0.0529 \log_{10} \dfrac{d_{ab}}{GMR_N}} I_a$$

Since $DN2$ and GMR_N are almost equal, if R_N is near zero, the neutral (or shield) current (I_{X1}) almost equals the phase current (I_a). Higher neutral resistances actually reduce positive-sequence resistances.

Significant effects on positive and zero-sequence impedances include:

- *Cable separation* — Larger separations increase Z_1; spacing does not affect Z_0. Triplex cables have the lowest positive-sequence impedance.

- *Conductor size* — Larger conductors have much less resistance; reactance drops somewhat with increasing size.

- *Neutral/shield resistance* — Increasing the neutral resistance increases the reactive portion of the positive and zero-sequence impedances. Beyond a certain point, increasing neutral resistances decreases the resistive portion of Z_1 and Z_0.

- *Other cables or ground wires* — Adding another grounded wire nearby has similar impacts to lowering sheath resistances. Zero-sequence resistance and reactance usually drop. Positive-sequence reactance is likely to decrease, but positive-sequence resistance may increase.

Figure 3.7 and Figure 3.8 show the impact of the most significant variables on impedances for three-phase and single-phase circuits. None of the following significantly impacts either the positive or zero-sequence impedances: insulation thickness, insulation type, depth of burial, and earth resistivity.

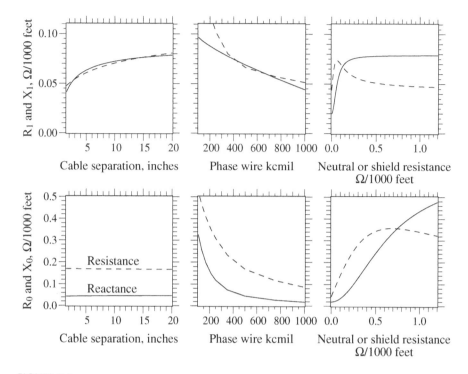

FIGURE 3.7
Effect of various parameters on the positive-sequence (top row) and zero-sequence impedances (bottom row) with a base case having 500-kcmil aluminum cables with 1/3 neutrals, 220-mil insulation, a horizontal configuration with 7.5 in. between cables, and ρ = 100 Ω-m.

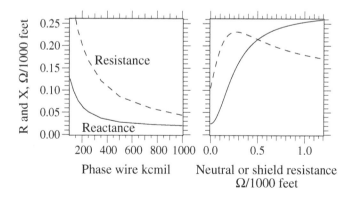

FIGURE 3.8
Resistance and reactance of a single-phase cable ($R = R_0 = R_1$ and $X = X_0 = X_1$) as the size of the cable and neutral varies with a base case having a 4/0 aluminum cable with a full neutral, 220-mil insulation, and ρ = 100 Ω-m.

TABLE 3.13

Loop Impedances of Single-Phase Concentric-Neutral
Aluminum Cables

Conductor Size	Full Neutral			1/3 Neutral		
	Neutral	R	X	Neutral	R	X
2	10#14	0.4608	0.1857			
1	13#14	0.3932	0.1517			
1/0	16#14	0.3342	0.1259	6#14	0.3154	0.2295
2/0	13#12	0.2793	0.0974	7#14	0.2784	0.2148
3/0	16#12	0.2342	0.0779	9#14	0.2537	0.1884
4/0	13#10	0.1931	0.0613	11#14	0.2305	0.1645
250	16#10	0.1638	0.0493	13#14	0.2143	0.1444
350	20#10	0.1245	0.0387	18#14	0.1818	0.1092
500				16#12	0.1447	0.0726
750				15#10	0.1067	0.0462
1000				20#10	0.0831	0.0343

Note: Impedances, $\Omega/1000$ ft ($\times$ 5.28 for $\Omega/$mi or $\times$ 3.28 for $\Omega/$km). Conductor temperature = 90°C, neutral temperature = 80°C, 15-kV class, 220-mil insulation, ρ = 100 Ω-m. For the neutral, 10#14 means 10 strands of 14-gage wire.

3.4.3 Impedance Tables

This section contains tables of several common cable configurations found on distribution circuits. All values are for a multigrounded circuit. Many other cable configurations are possible, with widely varying impedances. For PILC cables, refer to impedances in the Westinghouse (1950) T&D book. For additional three-phase power cable configurations, refer to the IEEE Red Book (IEEE Std. 141-1993), St. Pierre (2001), or Southwire Company (1997).

3.4.4 Capacitance

Cables have significant capacitance, much more than overhead lines. A single-conductor cable has a capacitance given by:

$$C = \frac{0.00736\varepsilon}{\log_{10}\left(\dfrac{D}{d}\right)}$$

where
  C = capacitance, μF/1000 ft
  ε = dielectric constant (2.3 for XLPE, 3 for EPR, see Table 3.4 for others)
  d = inside diameter of the insulation, mil (or other distance unit)
  D = outside diameter of the insulation in the same units as d

TABLE 3.14

Impedances of Three-Phase Circuits Made of Three Single-Conductor Concentric-Neutral Aluminum Cables

Conductor Size	Neutral Size	R_1	X_1	R_0	X_0	R_S	X_S
Full Neutral							
2	10#14	0.3478	0.1005	0.5899	0.1642	0.4285	0.1217
1	13#14	0.2820	0.0950	0.4814	0.1166	0.3484	0.1022
1/0	16#14	0.2297	0.0906	0.3956	0.0895	0.2850	0.0902
2/0	13#12	0.1891	0.0848	0.3158	0.0660	0.2314	0.0785
3/0	16#12	0.1578	0.0789	0.2573	0.0523	0.1910	0.0701
4/0	13#10	0.1331	0.0720	0.2066	0.0423	0.1576	0.0621
250	16#10	0.1186	0.0651	0.1716	0.0356	0.1363	0.0553
350	20#10	0.0930	0.0560	0.1287	0.0294	0.1049	0.0471
1/3 Neutral							
1/0	6#14	0.2180	0.0959	0.5193	0.2854	0.3185	0.1591
2/0	7#14	0.1751	0.0930	0.4638	0.2415	0.2713	0.1425
3/0	9#14	0.1432	0.0896	0.4012	0.1787	0.2292	0.1193
4/0	11#14	0.1180	0.0861	0.3457	0.1375	0.1939	0.1032
250	13#14	0.1034	0.0833	0.3045	0.1103	0.1704	0.0923
350	18#14	0.0805	0.0774	0.2353	0.0740	0.1321	0.0762
500	16#12	0.0656	0.0693	0.1689	0.0468	0.1000	0.0618
750	15#10	0.0547	0.0584	0.1160	0.0312	0.0752	0.0494
1000	20#10	0.0478	0.0502	0.0876	0.0248	0.0611	0.0417

Note: Impedances, $\Omega/1000$ ft ($\times$ 5.28 for $\Omega/$mi or $\times$ 3.28 for $\Omega/$km). Resistances for a conductor temperature = 90°C and a neutral temperature = 80°C, 220-mil insulation (15 kV), $\rho = 100$ Ω-m. Flat spacing with a 7.5-in. separation between cables. For the neutral, 10#14 means 10 strands of 14-gage wire.

The vars provided by cable are

$$Q_{var} = 2\pi \cdot f \cdot C \cdot V_{LG,kV}^2$$

where
Q_{var} = var/1000 ft/phase
f = frequency, Hz
C = capacitance, $\mu F/1000$ ft
$V_{LG,kV}$ = line-to-ground voltage, kV

Table 3.17 shows capacitance values and reactive power produced by cables for typical cables. The table results are for XLPE cable with a dielectric constant (ε) of 2.3. For other insulation, both the capacitance and the reactive power scale linearly. For example, for EPR with $\varepsilon = 3$, multiply the values in Table 3.17 by 1.3 (3/2.3 = 1.3).

TABLE 3.15

Impedances of Single-Conductor Aluminum Power Cables with
Copper Tape Shields

Conductor Size	R_1	X_1	R_0	X_0	R_S	X_S
Flat spacing with a 7.5-in. separation						
2	0.3399	0.1029	0.6484	0.4088	0.4427	0.2049
1	0.2710	0.0990	0.5808	0.3931	0.3743	0.1971
1/0	0.2161	0.0964	0.5268	0.3790	0.3196	0.1906
2/0	0.1721	0.0937	0.4833	0.3653	0.2759	0.1842
3/0	0.1382	0.0911	0.4494	0.3493	0.2419	0.1771
4/0	0.1113	0.0883	0.4217	0.3314	0.2148	0.1693
250	0.0955	0.0861	0.4037	0.3103	0.1982	0.1609
350	0.0696	0.0822	0.3734	0.2827	0.1709	0.1490
500	0.0508	0.0781	0.3483	0.2557	0.1499	0.1373
750	0.0369	0.0732	0.3220	0.2185	0.1319	0.1216
1000	0.0290	0.0698	0.3018	0.1915	0.1200	0.1104
Triplex						
2	0.3345	0.0531	0.7027	0.4244	0.4573	0.1769
1	0.2655	0.0501	0.6330	0.4060	0.3880	0.1687
1/0	0.2105	0.0483	0.5767	0.3893	0.3326	0.1620
2/0	0.1666	0.0465	0.5310	0.3734	0.2880	0.1554
3/0	0.1326	0.0448	0.4944	0.3550	0.2532	0.1482
4/0	0.1056	0.0432	0.4636	0.3346	0.2249	0.1403
250	0.0896	0.0424	0.4418	0.3109	0.2070	0.1319
350	0.0637	0.0403	0.4067	0.2807	0.1780	0.1204
500	0.0447	0.0381	0.3769	0.2518	0.1554	0.1093
750	0.0308	0.0359	0.3443	0.2129	0.1353	0.0949
1000	0.0228	0.0348	0.3197	0.1853	0.1218	0.0850

Note: Impedances, $\Omega/1000$ ft ($\times 5.28$ for Ω/mi or $\times 3.28$ for Ω/km). Resistances for a conductor temperature = 90°C and a shield temperature
= 50°C, 220-mil insulation (15 kV), $\rho = 100$ Ω-m, 5-mil copper tape
shield with a lap of 20%.

3.5   Ampacity

A cable's ampacity is the maximum continuous current rating of the cable.
We should realize that while we may derive one number, say 480 A, for
ampacity during normal operations for a given conductor, there is nothing
magic about 480 A. The cable will not burst into flames at 481 A; the 480 A
is simply a design number. We don't want to exceed that current during
normal operations.

The insulation temperature is normally the limiting factor. By operating
below the ampacity of a given cable, we keep the cable insulation below its

TABLE 3.16

Impedances of Single-Conductor Copper Power Cables

Conductor Size	R_1	X_1	R_0	X_0	R_S	X_S
Flat spacing with a 7.5-in. separation						
2	0.2083	0.1029	0.5108	0.4401	0.3092	0.2153
1	0.1671	0.0991	0.4718	0.4267	0.2687	0.2083
1/0	0.1334	0.0965	0.4405	0.4115	0.2358	0.2015
2/0	0.1082	0.0938	0.4171	0.3967	0.2112	0.1948
3/0	0.0871	0.0911	0.3975	0.3794	0.1906	0.1872
4/0	0.0705	0.0884	0.3816	0.3626	0.1742	0.1798
250	0.0607	0.0862	0.3719	0.3471	0.1644	0.1732
350	0.0461	0.0823	0.3558	0.3181	0.1493	0.1609
500	0.0352	0.0782	0.3411	0.2891	0.1372	0.1485
750	0.0272	0.0732	0.3241	0.2490	0.1261	0.1318
1000	0.0234	0.0699	0.3104	0.2196	0.1191	0.1198
Triplex						
2	0.2032	0.0508	0.5707	0.4642	0.3257	0.1886
1	0.1619	0.0477	0.5301	0.4480	0.2846	0.1811
1/0	0.1281	0.0460	0.4966	0.4295	0.2509	0.1738
2/0	0.1028	0.0442	0.4709	0.4116	0.2255	0.1667
3/0	0.0816	0.0426	0.4485	0.3910	0.2039	0.1587
4/0	0.0649	0.0409	0.4299	0.3713	0.1866	0.1510
250	0.0551	0.0398	0.4175	0.3532	0.1759	0.1442
350	0.0403	0.0377	0.3962	0.3202	0.1589	0.1319
500	0.0292	0.0355	0.3765	0.2882	0.1450	0.1197
750	0.0211	0.0333	0.3524	0.2450	0.1315	0.1039
1000	0.0173	0.0322	0.3336	0.2142	0.1227	0.0929

Note: Impedances, $\Omega/1000$ ft ($\times$ 5.28 for Ω/mi or $\times$ 3.28 for Ω/km). Resistances for a conductor temperature = 90°C and a shield temperature = 50°C, 220-mil insulation (15 kV), $\rho = 100$ Ω-m, 5-mil copper tape shield with a lap of 20%.

TABLE 3.17

Cable Capacitance for Common Cable Sizes and Voltages

	Capacitance, μF/1000 ft				Reactive power, kvar/1000 ft			
					12.5 kV	12.5 kV	25 kV	34.5 kV
Size	175 mil	220 mil	260 mil	345 mil	175 mil	220 mil	260 mil	345 mil
2	0.0516	0.0441	0.0396	0.0333	1.01	0.862	3.09	4.98
1	0.0562	0.0479	0.0428	0.0358	1.1	0.936	3.35	5.35
1/0	0.0609	0.0516	0.046	0.0383	1.19	1.01	3.6	5.72
2/0	0.0655	0.0553	0.0492	0.0407	1.28	1.08	3.84	6.09
3/0	0.0712	0.0599	0.0531	0.0437	1.39	1.17	4.15	6.54
4/0	0.078	0.0654	0.0578	0.0473	1.52	1.28	4.52	7.08
250	0.0871	0.0727	0.064	0.0521	1.7	1.42	5.00	7.79
350	0.0995	0.0826	0.0725	0.0586	1.94	1.61	5.67	8.76
500	0.113	0.0934	0.0817	0.0656	2.21	1.83	6.38	9.81
750	0.135	0.111	0.0969	0.0772	2.65	2.18	7.57	11.5
1000	0.156	0.127	0.111	0.0875	3.04	2.49	8.64	13.1

Note: For XLPE cable with $\varepsilon = 2.3$.

recommended maximum temperature. Cross-linked polyethylene cables are rated for a maximum operating temperature of 90°C during normal operations. Operating cables above their ampacity increases the likelihood of premature failures: water trees may grow faster, thermal runaway-failures are more likely, and insulation strength may decrease. In addition to absolute temperature, thermal cycling also ages cable more quickly.

Ampacity most often limits the loading on a cable; rarely, voltage drop or flicker limits loadings. Relative to overhead lines, cables of a given size have lower impedance and lower ampacities. So cable circuits are much less likely than overhead circuits to be voltage-drop limited. Only very long cable runs on circuits with low primary voltages are voltage-drop limited. Ampacity is not the only consideration for cable selection; losses and stocking considerations should also factor into cable selection. Choosing the smallest cable that meets ampacity requirements has the lowest initial cost, but since the cable is running hotter, the cost over its life may not be optimal because of the losses. Also allow for load growth when selecting cables.

Ampacity calculations follow simple principles: the temperature at the conductor is a function of the heat generated in a cable (I^2R) and the amount of heat conducted away from the cable. We can model the thermal performance with a thermal circuit analogous to an electric circuit: heat is analogous to current; temperature to voltage; and thermal resistance to electrical resistance. Heat flow through a thermal resistance raises the temperature between the two sides of the thermal material. Higher resistance soils or insulations trap the heat and cause higher temperatures. Using the thermal equivalent of Ohm's law, the temperature difference is:

$$\Delta T = T_C - T_A = R_{TH}H = R_{TH}(I^2R)$$

where
  T_C = conductor temperature, °C
  T_A = ambient earth temperature, °C
  R_{TH} = total thermal resistance between the cable conductor and the air, thermal Ω-ft
  H = heat generated in the cable, W (= I^2R)
  I = electric current in the conductor, A
  R = electric resistance of the conductor, Ω/ft

Most ampacity tables and computer calculation routines are based on the classic paper by Neher and McGrath (1957). The original paper is an excellent reference. Ander's book (1998) provides a detailed discussion of cable ampacity calculations, including the Neher–McGrath method along with IEC's method that is very similar (IEC 287, 1982). Hand calculations or spreadsheet calculations of the Neher–McGrath equations are possible, but tiresome; while straightforward in principle, the calculations are very detailed. A review of the Neher–McGrath procedure — the inputs, the tech-

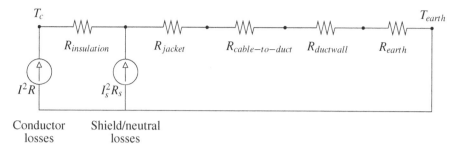

FIGURE 3.9
Thermal circuit model of a cable for ampacity calculations.

niques, the assumptions — provides a better understanding of ampacity calculations to better use computer ampacity calculations.

The Neher–McGrath procedure solves for the current in the equation above. Figure 3.9 shows a simplified model of the thermal circuit. The two main sources of heat within the cable are the I^2R losses in the phase conductor and the I^2R losses in the neutral or shield. The cable also has dielectric losses, but for distribution-class voltages, these are small enough that we can neglect them. The major thermal resistances are the insulation, the jacket, and the earth. If the cable system is in a duct, the air space within the duct and the duct walls adds thermal resistance. These thermal resistances are calculated from the thermal resistivities of the materials involved. For example, the thermal resistance of the insulation, jacket, and duct wall are all calculated with an equation of the following form:

$$R = 0.012\rho \log_{10}(D / d)$$

where
　　R = thermal resistance of the component, thermal Ω-ft
　　ρ = thermal resistivity of the component material, °C-cm/W
　　D = outside diameter of the component
　　d = inside diameter of the component

Thermal resistivity quantifies the insulating characteristics of a material. A material with $\rho = 1$°C-cm/W has a temperature rise of 1°C across two sides of a 1-cm³ cube for a flow of one watt of heat through the cube. As with electrical resistivity, the inverse of thermal resistivity is thermal conductivity. Table 3.18 shows resistivities commonly used for cable system components. The thermal resistance of a material quantifies the radial temperature rise from the center outward. One thermal Ω-ft has a radial temperature rise of 1°C for a heat flow of 1 W per ft of length (length along the conductor). Mixing of metric (SI) units with English units comes about for historical reasons.

TABLE 3.18

Thermal Resistivities of Common Components

Component	Thermal Resistivity, °C-cm/W
XLPE insulation	350
EPR insulation	500
Paper insulation	700
PE jackets	350
PVC jackets	500
Plastic ducts	480
Concrete	85
Thermal fill	60
Soil	90
Water	160
Air	4000

Sources: IEC 287, *Calculation of the Continuous Current Rating of Cables (100% Load Factor)*, 2nd ed., International Electrical Commission (IEC), 1982; Neher, J. H. and McGrath, M. H., "The Calculation of the Temperature Rise and Load Capability of Cable Systems," *AIEE Transactions*, vol. 76, pp. 752–64, October 1957.

TABLE 3.19

Ampacities of Single-Phase Circuits of Full-Neutral Aluminum Conductor Cables

Size	Direct Buried Load Factor		In Conduit Load Factor	
	100%	75%	100%	75%
2	187	201	146	153
1	209	225	162	170
1/0	233	252	180	188
2/0	260	282	200	210
3/0	290	316	223	234
4/0	325	356	249	262
250	359	395	276	291
350	424	469	326	345

Note: 90°C conductor temperature, 25°C ambient earth temperature, $\rho = 90$°C-cm/W.

The Neher–McGrath calculations also account for multiple cables, cables with cyclic daily load cycles, external heat sources, duct arrangements, and shield resistance and grounding variations.

Often, the easiest way to find ampacities for a given application is with ampacity tables. Table 3.19 and Table 3.20 show ampacities for common distribution configurations. Of the many sources of ampacity tables, the IEEE publishes the most exhaustive set of tables (IEEE Std. 835-1994). The National Electrical Code (NFPA 70, 1999) and manufacturer's publications (Okonite, 1990; Southwire Company, 1997) are also useful. Ampacity tables provide a

TABLE 3.20

Ampacities of Three-Phase Circuits Made
of Single-Conductor, One-Third Neutral
Aluminum Cables

	Direct Buried Load factor		In Conduit Load factor	
Size	100%	75%	100%	75%

Flat spacing (7.5-in. separation)

1/0	216	244	183	199
2/0	244	277	207	226
3/0	274	312	233	255
4/0	308	352	262	287
250	336	386	285	315
350	392	455	334	370
500	448	525	382	426
750	508	601	435	489
1000	556	664	478	541

Triplex

1/0	193	224	158	173
2/0	220	255	180	197
3/0	249	290	204	225
4/0	283	330	232	256
250	312	365	257	284
350	375	442	310	345
500	452	535	375	419
750	547	653	457	514
1000	630	756	529	598

Note: 90°C conductor temperature, 25°C ambient earth temperature, ρ = 90°C-cm/W.

good starting point for determining the ampacity of a specific cable application. When using tables, be careful that the assumptions match your particular situation; if not, ampacity results can be much different than expected.

Conductor temperature limits, sheath resistance, thermal resistivity of the
soil — these are some of the variables that most impact ampacity (see Figure
3.10). These and other effects are discussed in the next few paragraphs [see
also (CEA, 1982; NRECA RER Project 90-8, 1993) for more discussions].

Sheath resistance — On a three-phase circuit, the resistance of the sheath
(or shield or neutral) plays an important role in ampacity calculations.
Because a cable's phase conductor and sheath couple so tightly, current
through the phase induces a large voltage along the sheath. With the cable
sheath grounded periodically, circulating current flows to counter the
induced voltage. The circulating current is a function of the resistance of the
sheath. This circulating current leads to something counterintuitive: sheaths
with higher resistance have more ampacity. Higher resistance sheaths reduce
the circulating current and reduce the I^2R losses in the sheath. This effect is

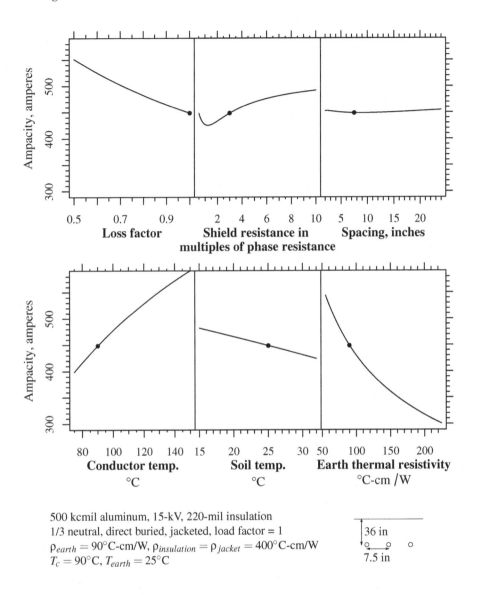

500 kcmil aluminum, 15-kV, 220-mil insulation
1/3 neutral, direct buried, jacketed, load factor = 1
$\rho_{earth} = 90°\text{C-cm/W}$, $\rho_{insulation} = \rho_{jacket} = 400°\text{C-cm/W}$
$T_c = 90°\text{C}$, $T_{earth} = 25°\text{C}$

36 in

7.5 in

FIGURE 3.10
Effect of variables on ampacity for an example cable.

most pronounced in larger conductors. Many ampacity tables assume that
cable sheaths are open circuited, this eliminates the sheath losses and
increases the ampacity. The open-circuit sheath values can be approximately
corrected to account for circulating currents (Okonite, 1990) by

$$k = \sqrt{\frac{I^2 R}{I_S^2 R_S + I^2 R}}$$

where

k = ampacity multiplier to account for sheath losses, i.e., $I_{\text{grounded sheath}} = k \cdot I_{\text{open sheath}}$

I = phase conductor current, A

I_S = sheath current, A

I^2R = phase conductor losses, W/unit of length

$I_S^2 R_S$ = sheath losses, W/unit of length

The sheath losses are a function of the resistance of the sheath and the mutual inductance between the sheath and other conductors. For a triangular configuration like triplex, the shield losses are

$$I_S^2 R_S = I^2 R_S \frac{X_M^2}{R_S^2 + X_M^2}$$

where

$$X_M = 2\pi f(0.1404)\log_{10}(2S / d_S)$$

and

X_M = mutual inductance of the sheath and another conductor, mΩ/1000 ft

R_S = resistance of the sheath, mΩ/1000 ft

f = frequency, Hz

S = spacing between the phase conductors, in.

d_S = mean diameter of the sheath, in.

For configurations other than triplex, see Southwire Company (1997) or Okonite (1990). Figure 3.11 shows how sheath losses vary with conductor size and with spacing. Spacing has a pronounced effect. Steel ducts can significantly increase heating from circulating currents. In fact, even nearby steel pipes can significantly reduce ampacity.

Spacings — Separating cables separates the heat sources. But at larger spacings, circulating currents are higher. Optimal spacings involve balancing these effects. For smaller cables, separating cables provides the best ampacity. For larger cables (with larger circulating currents), triplex or other tight spacing improves ampacity. For one-third neutral, aluminum cables, NRECA (1993) shows that a flat spacing with 7.5 in. between cables has better ampacity than triplex for conductors 500 kcmil and smaller. For copper cables, the threshold is lower: conductors larger than 4/0 have better ampacity with a triplex configuration.

Conductor temperature — If we allow a higher conductor temperature, we can operate a cable at higher current. If we know the ampacity for a given conductor temperature, at a different conductor temperature we can find the ampacity with the following approximation:

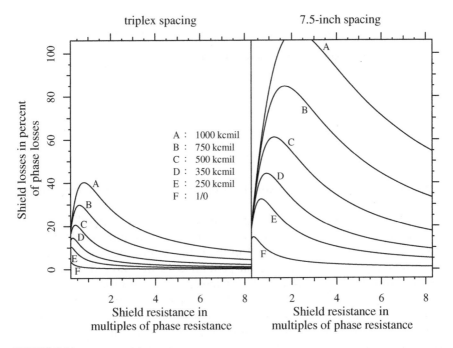

FIGURE 3.11
Shield losses as a function of shield resistance for aluminum cables (triplex configuration).

$$I' = I\sqrt{\frac{T_C' - T_A'}{T_C - T_A}\frac{228.1 + T_C}{228.1 + T_C'}} \qquad \text{(Aluminum conductor)}$$

$$I' = I\sqrt{\frac{T_C' - T_A'}{T_C - T_A}\frac{234.5 + T_C}{234.5 + T_C'}} \qquad \text{(Copper conductor)}$$

where

I' = ampacity at a conductor temperature of T_C' and an ambient earth temperature T_A'

I = ampacity at a conductor temperature of T_C and an ambient earth temperature T_A (all temperatures are in °C)

We can use these equations to find emergency ampacity ratings of cables. In an emergency, XLPE can be operated to 130°C. Some EPR cables can be operated to 140°C (MV-105 cables). ICEA standards allow emergency overload for 100 hours per year with five such periods over the life of the cable. Polyethylene cables, including HMWPE, have little overload capability. Their maximum recommended emergency temperature is 95°C. Table 3.21 shows common ampacity multipliers; these are valid for both copper and aluminum conductors within the accuracy shown. We can also use the

TABLE 3.21

Common Ampacity Rating Conversions (with $T_A = 25°C$)

Original Temperature, °C	New Temperature, °C	Ampacity Multiplier
75	95	1.15
90	75	0.90
90	105	1.08
90	130	1.20
105	140	1.14

appropriate temperature-adjustment equation to adjust for different ambient earth temperatures.

Loss factor — The earth has a high thermal storage capability; it takes considerable time to heat (or cool) the soil surrounding the cable. Close to the cable, the peak heat generated in the cable determines the temperature drop; farther out, the average heat generated in the cable determines the temperature drop. As discussed in Chapter 5, we normally account for losses using the loss factor, which is the average losses divided by the peak losses. Since this number is not normally available, we find the loss factor from the load factor (the load factor is the average load divided by the peak load). Assuming a 100% load factor (continuous current) is most conservative but can lead to a cable that is larger than necessary. We should try to err on the high side when estimating the load factor. A 75% load factor is commonly used.

Conduits — The air space in conduits or ducts significantly reduces ampacity. The air insulation barrier traps more heat in the cable. Direct-buried cables may have 10 to 25% higher ampacities. Although the less air the better, there is little practical difference in the thermal performance between the sizes of ducts commonly used. Concrete duct banks have roughly the same thermal performance as direct-buried conduits (concrete is more consistent and less prone to moisture fluctuations).

Soil thermal resistivity and temperature — Soils with lower thermal resistivity more readily conduct heat away from cables. Moisture is an important component, moist soil has lower thermal resistivity (see Figure 3.12). Dense soil normally has better conductivity. More so than any other single factor, soil resistivity impacts the conductor's temperature and the cable's ampacity. A resistivity of 90°C-cm/W is often assumed for ampacity calculations. This number is conservative enough for many areas, but if soil resistivities are higher, cable temperatures can be much higher than expected. For common soils, Table 3.22 shows typical ranges of thermal resistivities. At typical installation depths, resistivity varies significantly with season as moisture content changes. Unfortunately in many locations, just when we need ampacity the most — during peak load in the summer — the soil is close to its hottest and driest. Seasonal changes can be significant, but daily

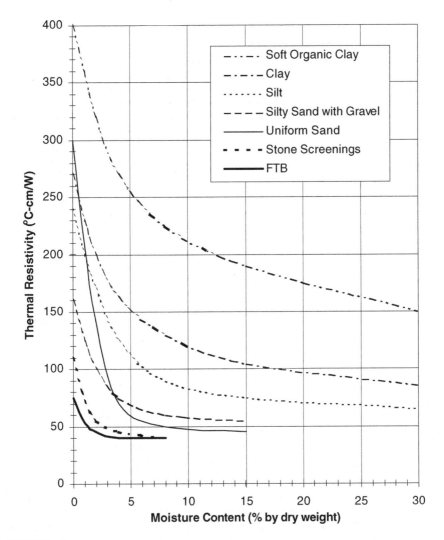

FIGURE 3.12
Effect of moisture on the thermal resistivity of various soils. (Copyright © 1997. Electric Power Research Institute. TR-108919. *Soil Thermal Properties Manual for Underground Power Transmission.* Reprinted with permission.)

changes are not; soil temperature changes lag air temperature changes by 2 to 4 weeks.

The depth of burial can affect ampacity. With a constant resistivity and soil temperature, deeper burial decreases ampacity. But deeper, the soil tends to have lower temperature, more moisture, and soil is more stable seasonally. To go deep enough to take advantage of this is not cost effective though.

For areas with poor soil (high clay content in a dry area, for example), one of several thermal backfills can give good performance, with stable

TABLE 3.22

Typical Thermal Resistivities of Common Soils

USCS	Soil	Dry Density (g/cm³)	Range of Moisture Contents (%) Above Water Table	Saturated Moisture Content (%)	Thermal Resistivity (°C-cm/W) Wet–Dry
GW	well graded gravel	2.1	3–8	10	40–120
GP	poor graded gravel	1.9	2–6	15	45–190
GM	silty gravel	2.0	4–9	12	50–140
GC	clayey gravel	1.9	5–12	15	55–150
SW	well graded sand	1.8	4–12	18	40–130
SP	uniform sand	1.6	2–8	25	45–300
SM	silty sand	1.7	6–16	20	55–170
SC	clayey sand	1.6	8–18	25	60–180
ML	Silt	1.5	8–24	30	65–240
CL	silty clay	1.6	10–22	25	70–210
OL	organic silt	1.2	15–35	45	90–350
MH	micaceous silt	1.3	12–30	40	75–300
CH	clay	1.3	20–35	40	85–270
OH	soft organic clay	0.9	30–70	75	110–400
Pt	silty peat	0.4			150–600+

resistivities below 60°C-cm/W even when moisture content drops below one percent.

Earth interface temperature — Because soil conductivity depends on moisture, the temperature at the interface between the cable or duct and the soil is important. Unfortunately, heat tends to push moisture away. High interface temperatures can dry out the surrounding soil, which further increases the soil's thermal resistivity. Soil drying can lead to a runaway situation; hotter cable temperatures dry the soil more, raising the cable temperature more and so on. Some soils, especially clay, shrink significantly as it dries; the soil can pull away from the cable, leaving an insulating air layer. Thermal runaway can lead to immediate failure. Direct-buried cables are the most susceptible; ducts provide enough of a barrier that temperature is reduced by the time it reaches the soil.

Depending on the soil drying characteristics in an area, we may decide to limit earth interface temperatures. Limiting earth interface temperatures to 50 to 60°C reduces the risk of thermal runaway. But doing this also significantly decreases the ampacity of direct-buried cable to about that of cables in conduit. In fact, using the conduit ampacity values is a good approximation for the limits needed to keep interface temperatures in the 50 to 60°C range.

Current unbalance — Almost every ampacity table (including those in this section) assumes balanced, three-phase currents. On multigrounded distribution systems, this assumption is rarely true. An ampacity of 100 A means a limit of 100 A on each conductor. Unbalance restricts the power a three-

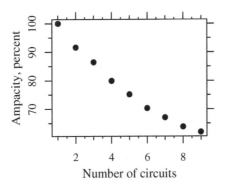

FIGURE 3.13
Ampacity reduction with multiple cable circuits in a duct bank (15 kV, aluminum, 500 kcmil, tape shield power cables, triplex configuration).

phase cable circuit can carry ($I_A = I_B = I_C = 100$ A carries more power than $I_A = 100$ A, $I_B = I_C = 70$ A). In addition, the unbalanced return current may increase the heating in the cable carrying the highest current. It may or it may not; it depends on phase relationships and the phase angle of the unbalanced current. If the unbalances are just right, the unbalanced return current can significantly increase the neutral current on the most heavily loaded phases. Unbalance also depends on the placement of the cables. In a flat configuration, the middle cable is the most limiting because the outer two cables heat the middle cable.

Just as higher sheath resistances reduce circulating currents, higher sheath resistances reduce unbalance currents in the sheath. Higher sheath resistances force more of the unbalanced current to return in the earth. The heat generated in the sheath from unbalance current also decreases with increasing sheath resistance (except for very low sheath resistances, where the sheath has less resistance than the phase conductor).

System voltage and insulation thickness — Neither significantly impacts the ampacity of distribution cables. Ampacity stays constant with voltage; 5-kV cables have roughly the same ampacity as 35-kV cables. At higher voltages, insulation is thicker, but this rise in the thermal resistance of the insulation reduces the ampacity just slightly. Higher operating voltages also cause higher dielectric losses, but again, the effect is small (it is more noticeable with EPR cable).

Number of cables — Cables in parallel heat each other, which restricts ampacity. Figure 3.13 shows an example for triplex power cables in duct banks.

Cable crossings and other hotspots — Tests have found that cable crossings can produce significant hotspots (Koch, 2001). Other hotspots can occur in locations where cables are paralleled for a short distance like taps to pad-mounted transformers or other gear. Differences in surface covering (such as asphalt roads) can also produce hot spots. Anders and Brakelmann (1999a, 1999b) provide an extension to the Neher–McGrath model that includes the effects of cable crossings at different angles. They conclude: "the derating of

3 to 5% used by some utilities may be insufficient, especially for cables with smaller conductors."

Riser poles — Cables on a riser pole require special attention. The protective vertical conduit traps air, and the sun adds external heating. Hartlein and Black (1983) tested a specific riser configuration and developed an analytical model. They concluded that the size of the riser and the amount of venting were important. Large diameter risers vented at both ends are the best. With three cables in one riser, they found that the riser portion of the circuit limits the ampacity. This is especially important in substation exit cables and their riser poles. In a riser pole application, ampacity does not increase for lower load factors; a cable heats up much faster in the air than when buried in the ground (the air has little thermal storage). NRECA (1993) concluded that properly vented risers do not need to be derated, given that venting can increase ampacity between 10 and 25%. If risers are not vented, then the riser becomes the limiting factor. Additional work in this area has been done by Cress (1991) (tests and modeling for submarine cables in riser poles) and Anders (1996) (an updated analytical model).

3.6   Fault Withstand Capability

Short-circuit currents through a conductor's resistance generates tremendous heat. All cable between the source and the fault is subjected to the same phase current. For cables, the weakest link is the insulation; both XLPE and EPR have a short-duration upper temperature limit of 250°C. The short-circuit current injects energy as a function of the fault duration multiplied by the square of the current.

For aluminum conductors and XLPE or EPR insulation, the maximum allowable time-current characteristic is given by

$$I^2 t = (48.4A)^2$$

where
   I = fault current, A
   t = fault duration, sec
   A = cross-sectional area of the conductor, kcmil

This assumes an upper temperature limit of 250°C and a 90°C starting temperature. For copper, the upper limit is defined by

$$I^2 t = (72.2A)^2$$

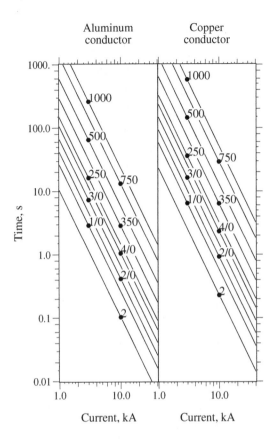

FIGURE 3.14
Short-circuit limit of cables with EPR or XLPE insulation.

We can plot these curves along with the time-current characteristics of the protecting relay, fuse, or recloser to ensure that the protective devices protect our cables.

Damage to the shield or the neutral is more likely than damage to the phase conductor. During a ground fault, the sheath may conduct almost as much current as the phase conductor, and the sheath is normally smaller. With a one-third neutral, the cable neutral's I^2t withstand is approximately 2.5 times less than the values for the phase conductor indicated in Figure 3.14 (this assumes a 65°C starting temperature). Having more resistance, a tape shield is even more vulnerable. A tape shield has a limiting time-current characteristic of

$$I^2t = (z \cdot A)^2$$

where z is 79.1 for sheaths of copper, 58.2 for bronze, 39.2 for zinc, 23.7 for copper-nickel, and 15 for lead [with a 65°C starting temperature and an

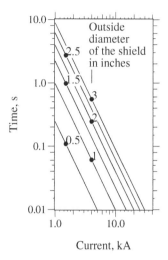

FIGURE 3.15
Short-circuit insulation limit of copper tape shields based on outside diameter (starting temperature is 65°C, final temperature is 250°C, 20% lap on the shield).

upper limit of 250°C; using data from (Kerite Company)]. Figure 3.15 shows withstand characteristics for a 5-mil copper tape shield. The characteristic changes with cable size because larger diameter cables have a shield with a larger circumference and more cross-sectional area. If a given fault current lasts longer than five times the insulation withstand characteristic (at 250°C), the shield reaches its melting point.

In the vicinity of the fault, the fault current can cause considerably more damage to the shield or neutral. With a concentric neutral, the fault current may only flow on a few strands of the conductor until the cable has a grounding point where the strands are tied together. Excessive temperatures can damage the insulation shield, the insulation, and the jacket. In addition, the temperature may reach levels that melt the neutral strands. A tape shield can suffer similar effects: where tape layers overlap, oxidation can build up between tape layers, which insulate the layers from each other. This can restrict the fault current to a smaller portion of the shield. Additionally, where the fault arc attaches, the arc injects considerable heat into the shield or neutral, causing further damage at the failure point. Some additional damage at the fault location must be tolerated, but the arc can burn one or more neutral strands several feet back toward the source.

Martin et al. (1974) reported that longitudinally corrugated sheaths perform better than wire or tape shields for high fault currents. They also reported that a semiconducting jacket helped spread the fault current to the sheaths of other cables (the semiconducting material breaks down).

Pay special attention to substation exit cables in areas with high fault currents (especially since exit cables are critical for circuit reliability). During a close fault, where currents are high, a reduced neutral or tape shield is most prone to damage.

3.7 Cable Reliability

3.7.1 Water Trees

The most common failure cause of solid-dielectric cables has been *water treeing*. Water trees develop over a period of many years and accelerate the failure of solid dielectric cables. Excessive treeing has led to the premature failure of many polyethylene cables. Cable insulation can tree two ways:

- *Electrical trees* — These hollow tubes develop from high electrical stress; this stress creates partial discharges that eat away at the insulation. Once initiated, electrical trees can grow fast, failing cable within hours or days.

- *Water trees* — Water trees are small discrete voids separated by insulation. Water trees develop slowly, growing over a period of months or years. Much less electrical stress is needed to cause water trees. Water trees actually look more like fans, blooms, or bushes whereas electrical trees look more like jagged branched trees. As its name indicates, water trees need moisture to grow; water that enters the dielectric accumulates in specific areas (noncrystalline regions) and causes localized degradation. Voids, contaminants, temperature, and voltage stress — all influence the rate of growth.

The formation of water trees does not necessarily mean the cable will fail. A water tree can even bridge the entire dielectric without immediate failure. Failure occurs when a water tree converts to an electrical tree. One explanation of the initiation of electrical trees is from charges trapped in the cable insulation. In Thue's words (1999), "they can literally bore a tunnel from one void or contaminant to the next." Impulses and dc voltage (in a hi-pot test) can trigger electrical treeing in a cable that is heavily water treed.

The growth rate of water trees tends to reduce with time; as trees fan out, the electrical stress on the tree reduces. Trees that grow from contaminants near the boundary of the conductor shield are most likely to keep growing. These are "vented" trees. Bow-tie trees (those that originate inside the cable) tend to grow to a critical length and then stop growing.

The electrical breakdown strength of aged cable has variation, a variation that has a skewed probability distribution. Weibull or lognormal distributions are often used to characterize this probability and predict future failure probabilities.

Polyethylene insulation systems have been plagued by early failures caused by water trees. Early XLPE and especially HMWPE had increasing failure rates that have led utilities to replace large quantities of cable. By most accounts, polyethylene-based insulation systems have become much more resistant to water treeing and more reliable for many reasons (Dudas, 1994; EPRI 1001894, 2001; Thue, 1999):

- *Extruded semiconducting shields* — Rather than taped conductor and insulation shields, manufacturers extrude both semiconducting shields as they are extruding the insulation. This one-pass extrusion provides a continuous, smooth interface. The most dangerous water trees are those that initiate from imperfections at the interface between the insulation and the semiconducting shield. Reducing these imperfections reduces treeing.

- *Cleaner insulation* — AEIC specifications for the allowable number and size of contaminants and protrusions have steadily improved. Both XLPE compound manufacturers and cable manufacturers have reduced contaminants by improving their production and handling processes.

- *Fewer voids* — Dry curing reduces the number and size of voids in the cable. Steam-cured cables pass through a long vulcanizing tube filled with 205°C steam pressurized at 20 atm. Cables cured with steam have sizeable voids in the insulation. Instead of steam, dry curing uses nitrogen gas pressurized to 10 atm; an electrically heated tube radiates infrared energy that heats the cable. Dry curing has voids, but these voids have volumes 10 to 100 times less than with steam curing.

- *Tree-retardant formulations* — Tree-retardant formulations of XLPE perform much better in accelerated aging tests, tests of field-aged cables, and also in field experience.

EPR insulation has proven to be naturally water tree resistant; EPR cables have performed well in service since the 1970s. EPR insulation can and does have water trees, but they tend to be smaller. EPR cable systems have also improved by having cleaner insulation compounds, jackets, and extruded semiconductor shields.

Several accelerated aging tests have been devised to predict the performance of insulation systems. The tests use one of two main methods to quantify performance: (1) loss of insulation strength or (2) time to failure. In accelerated aging, testers normally submerge cables in water, operate the cables at a continuous overvoltage, and possibly subject the cables to thermal cycling. The accelerated water treeing test (AWTT) is a protocol that measures the loss of insulation strength of a set of samples during one year of testing (ANSI/ICEA S-94-649-2000, 2000). The wet aging as part of this test includes application of three times rated voltage and current sufficient to heat the water to 60°C. In another common test protocol, the accelerated cable life test (ACLT), cables are submerged in water, water is injected into the conductor strands, cables are operated to (commonly) four times nominal voltage, and cables are brought to 90°C for eight hours each day. The cables are operated to failure. Brown (1991) reported that under such a test, XLPE and TR-XLPE cables had geometric mean failure times of 53 and 161 days, respectively. Two EPR constructions did not fail after 597 days of testing. Because EPR and XLPE age differently depending on the type of stress, EPR can come out better or worse than TR-XLPE, depending on the test conditions. There is no consensus on the best accelerated-aging test. Normally such tests are used to compare two types of cable constructions. Bernstein concludes, "... there is still no acceptable means of relating service and laboratory aging to 'remaining life'" (EPRI 1000273, 2000).

Even without voltage, XLPE cable left outdoors can age. EPRI found that XLPE cables left in the Texas sun for 10 years lost over 25% of their ac insulation strength (EPRI 1001389, 2002). These researchers speculate that heating from the sun led to a loss of peroxide decomposition by-products, which is known to result in loss of insulation strength.

Since water promotes water treeing, a few utilities use different forms of water blocking (Powers, 1993). Water trees grow faster when water enters the insulation from both sides: into the conductor strands and through the cable sheath. The most common water-protection method is a filled strand conductor; moisture movement or migration is minimized by the filling, which can be a semiconducting or an insulating filler. Another variation uses water absorbing powders; as the powder absorbs water it turns to a gel that blocks further water movement. An industry standard water blocking test is provided (ICEA Publication T-31-610, 1994; ICEA Publication T-34-664, 1996). In addition to reducing the growth and initiation of water trees, a strand-blocked conductor reduces corrosion of aluminum phase conductors. We can also use solid conductors to achieve the same effect (on smaller cables).

Another approach to dealing with water entry and treeing in existing cable is to use a silicone injection treatment (Nannery et al., 1989). After injection into the stranded conductor, the silicone diffuses out through the conductor shield and into the insulation. The silicone fills water-tree voids and reacts with water such that it dries the cable. This increases the dielectric strength and helps prevent further treeing and loss of life.

Another way to increase the reliability is to increase the insulation thickness. As an example, the maximum electrical stress in a cable with an insulation thickness of 220 mil (1 mil = 0.001 in. = 0.00254 cm) is 14% lower than a 175-mil cable (Mackevich, 1988).

Utilities and manufacturers have taken steps to reduce the likelihood of cable degradation. Table 3.23 shows trends in cable specifications for underground residential cable. Tree-retardant insulation and smooth semiconductor shields, jackets and filled conductors, and dry curing and triple extrusion are features specified by utilities to improve reliability.

TABLE 3.23

Trends in URD Cable Specifications

Characteristic	1983	1988	1993	1998
XLPE insulation	84	52	20	0
TR-XLPE insulation		36	52	68
EPR insulation	12	12	28	32
Protective jacket	64	80	92	93[a]
Filled strand conductor	4	32	60	68
Dry cure for XLPE and TR-XLPE		24	56	52
Triple extrusion		44	64	67[a]
Supersmooth semicon shields		0	44	56
Bare copper neutrals		72	84	

Note: Percentage of the 25 largest investor-owned utilities in the U.S. that specify the given characteristic.

[a] Somewhat different data set: percentages from the top 45 largest investor-owned utilities.

Sources: Dudas, J. H., "Technical Trends in Medium Voltage URD Cable Specifications," *IEEE Electrical Insulation Magazine*, vol. 10, no. 2, pp. 7–16, March/April 1994; Dudas, J. H. and Cochran, W. H., "Technical Advances in the Underground Medium-Voltage Cable Specification of the Largest Investor-Owned Utilities in the U.S.," *IEEE Electrical Insulation Magazine*, vol. 15, no. 6, pp. 29–36, November/December 1999.

Good lightning protection also reduces cable faults. This requires surge arresters at the riser pole and possibly arresters at the cable open point (depending on the voltage). Keep arrester lead lengths as short as possible. Surges are a known cause of dielectric failures. Surges that do not fail the insulation may cause aging. Accelerated aging tests have found that 15-kV XLPE cables tested with periodic surges applied with magnitudes of 40, 70, and 120 kV failed more often and earlier than samples that were not surged (EPRI EL-6902, 1990; EPRI TR-108405-V1, 1997; Hartlein et al., 1989; Hartlein et al., 1994). Very few of the failures occurred during the application of a surge; this follows industry observations that cables often fail after a thunderstorm, not during the storm.

Rather than continue patching, many utilities regularly replace cable. Program policies are done based on the number of failures (the most common approach), cable inspection, customer complaints, or cable testing. High-molecular weight polyethylene and older XLPE are the most likely candidates for replacement. Most commonly, utilities replace cable after two or three electrical failures within a given time period (see Table 3.24).

3.7.2 Other Failure Modes

Cable faults can be caused by several events including:

TABLE 3.24

Typical Cable Replacement Criteria

Replacement Criteria	Responses (n = 51)
One failure	2%
Two failures	31%
Three failures	41%
Four failures	4%
Five failures	6%
Based on evaluation procedures	16%

Source: Tyner, J. T., "Getting to the Bottom of UG Practices," *Transmission & Distribution World*, vol. 50, no. 7, pp. 44–56, July 1998.

- Dig-ins
- Cable failures
- Cable equipment failures — splices, elbows, terminations

Better public communications reduces dig-ins into cables. The most common way is with one phone number that can be used to coordinate marking of underground facilities before digging is done. Physical methods of reducing dig-ins include marker tape, surface markings, or concrete covers. Marker tape identifies cable. A few utilities use surface marking to permanently identify the location of underground facilities. Concrete covers above underground facilities physically block dig-ins.

Temporary faults are unusual in underground facilities. Faults are normally bolted, permanent short circuits. Reclosing will just do additional damage to the cable. Occasionally, animals or water will temporarily fault a piece of live-front equipment. Recurring temporary faults like these can be very difficult to find.

Another type of temporary, self-clearing fault can occur on a cable splice (Stringer and Kojovic, 2001). Figure 3.16 shows a typical waveform of an impending splice failure. This type of fault has some distinguishing characteristics: it self-clears in 1/4 cycle, the frequency of occurrence increases with time, and faults occur near the peak of the voltage. The author has observed this type of fault during monitoring (but never identified the culprit). This type of fault can occur in a cable splice following penetration of water into the splice. The water breaks down the insulation, then the arc energy melts the water and creates vapor at high pressure. Finally, the high-pressure vapor extinguishes the arc. The process can repeat when enough water accumulates again until the failure is permanent. This type of self-clearing fault can go unnoticed until it finally fails. The downside is that it causes a short-duration voltage sag that may affect sensitive equipment. Another problem, the fault may have enough current to blow a fuse; but since the fault self-clears, it can be much harder to find. Crews may just replace the fuse (successfully) and leave without replacing the damaged equipment.

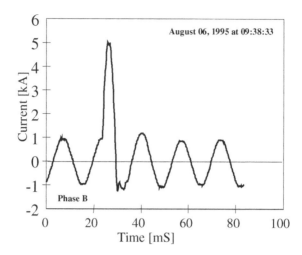

FIGURE 3.16
Self-clearing fault signature on an incipient cable-splice failure. (From Stringer, N. T. and Kojovic, L. A., "Prevention of Underground Cable Splice Failures," *IEEE Trans. Industry App.*, 37(1), 230-9, Jan./Feb. 2001. With permission. ©2001 IEEE.)

3.7.3 Failure Statistics

The annual failures of cables is on the order of 6 to 7 failures per 100 mi per year (3.7 to 4.3 failures per 100 km per year) according to survey data from the Association of Edison Illuminating Companies from 1965 through 1991 (Thue, 1999). Figure 3.17 shows cable failure data from a variety of sources; experience varies widely. Application, age, and type of cable markedly change the results. Utilities have experienced high failures of HMWPE, especially those that installed in the early 1970s. An EPRI database of 15 utilities showed a marked increase in failure rates for HMWPE cables with time (Stember et al., 1985). XLPE also shows a rise in failure rates with time, but not as dramatic (see Figure 3.18). The EPRI data showed failure rates increased faster with a higher voltage gradient on the dielectric for both HMWPE and XLPE.

Much of the failure data in Figure 3.17 is dominated by earlier polyethylene-based cable insulation technologies. Not as much data is available on the most commonly used insulation materials: TR-XLPE and EPR. The AEIC survey reported results in 1991 — both had fewer than 0.5 failures per 100 cable mi during that year. TR-XLPE results were better (0.2 vs. 0.4 failures/100 mi/year for EPR), but the installed base of TR-XLPE would have been newer than EPR at that time. Jacketed cable has had fewer failures than unjacketed cable as shown in Table 3.25.

Another consideration for underground circuits is the performance of connectors and other cable accessories. 200-A elbows have failed at high rates (and they tend to fail when switching under load) (Champion, 1986).

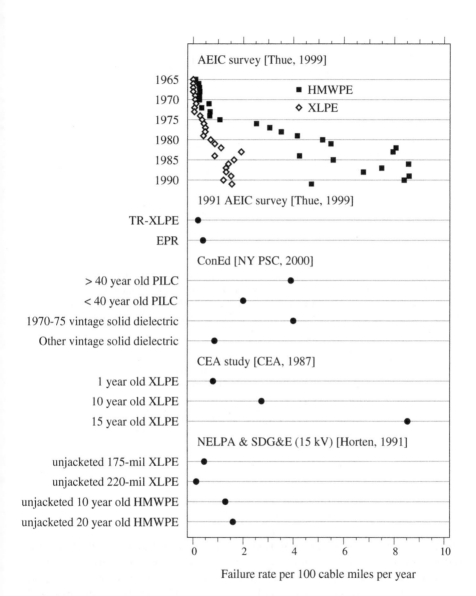

FIGURE 3.17
Cable failure rates found in different studies and surveys (in cable miles, not circuit miles). (Data from [CEA 117 D 295, 1987; Horton and Golberg, 1991; State of New York Department of Public Service, 2000; Thue, 1999].)

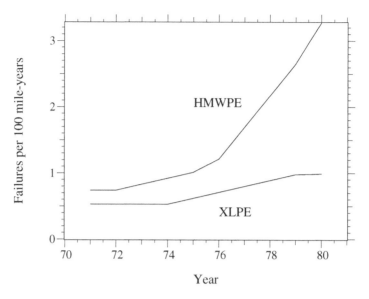

FIGURE 3.18
Cumulative service-time failure rates for HMWPE and XLPE cable. (From Stember, L. H., Epstein, M. M., Gaines, G. V., Derringer, G. C., and Thomas, R. E., "Analysis of Field Failure Data on HMWPE- and XLPE-Insulated High-Voltage Distribution Cable," *IEEE Trans. Power Apparatus Sy.*, PAS-104(8), 1979-85, August 1985. With permission. ©1985 IEEE.)

TABLE 3.25

Comparison of the Median of the Average
Yearly Failure Rates of XLPE Found by AEIC
from 1983 to 1991

Configuration	Failures per 100 Cable miles/year
No jacket	3.1
Jacketed	0.2
Direct buried	2.6
Duct	0.2

Source: Thue, W. A., *Electrical Power Cable Engineering*, Marcel Dekker, New York, 1999.

One important factor is that the type of splice should be correctly matched with the type of cable (Mackevich, 1988). Table 3.26 shows annual failure rates for some common underground components that were developed based on data from the Northwest Underground Distribution Committee of the Northwest Electric Light and Power Association (Horton and Golberg, 1990; Horton and Golberg, 1991). Table 3.27 shows failure rates of splices for New York City.

An EPRI review of separable connector reliability found mixed results (EPRI 1001732, 2002). Most utilities do not track these failures. One utility that did keep records found that failure rates of separable connectors ranged

TABLE 3.26

Annual Underground-Component Failure Rates

Component	Annual Failure Rate, %
Load-break elbows	0.009t
15-kV molded rubber splices	0.31
25-kV molded rubber splices	0.18
35-kV molded rubber splices	0.25
Single-phase padmounted transformers	0.3

Note: t is the age of the elbow in years.

Sources: Horton, W. F. and Golberg, S., "The Failure Rates of Underground Distribution System Components," Proceedings of the Twenty-Second Annual North American Power Symposium, 1990; Horton, W. F. and Golberg, S., "Determination of Failure Rates of Underground Distribution System Components from Historical Data," IEEE/PES Transmission and Distribution Conference, 1991.

TABLE 3.27

Underground Network Component Failure Rates in New York City (Con Edison)

Component	Annual Failure Rate, %
Splices connecting paper to solid cables (stop joints)	1.20
Splices connecting similar cables (straight joints)	0.51
Network transformers	0.58

Source: State of New York Department of Public Service, "A Report on Consolidated Edison's July 1999 System Outages," March 2000.

from 0.1 to 0.4% annually. Of these failures, an estimated 3 to 20% are from overheating. They also suggested that thermal monitoring is a good practice, but effectiveness is limited because the monitoring is often done when the loadings and temperatures are well below their peak.

3.8 Cable Testing

A common approach to test cable and determine insulation integrity is to use a hi-pot test. In a hi-pot test, a dc voltage is applied for 5 to 15 min. IEEE-400 specifies that the hi-pot voltage for a 15-kV class cable is 56 kV for an acceptance test and 46 kV for a maintenance test (ANSI/IEEE Std. 400-1980). Other industry standard tests are given in (AEIC CS5-94, 1994; AEIC CS6-96, 1996; ICEA S-66-524, 1988). High-pot testing is a brute-force test; imminent failures are detected, but the amount of deterioration due to aging is not quantified (it is a go/no–go test).

The dc test is controversial — some evidence has shown that hi-pot testing may damage XLPE cable (Mercier and Ticker, 1998). EPRI work has shown that dc testing accelerates treeing (EPRI TR-101245, 1993; EPRI TR-101245-V2, 1995). For hi-pot testing of 15-kV, 100% insulation (175-mil, 4.445-mm) XLPE cable, EPRI recommended:

- Do not do testing at 40 kV (228 V/mil) on cables that are aged (especially those that failed once in service and then are spliced). Above 300 V/mil, deterioration was predominant.
- New cable can be tested at the factory at 70 kV. No effect on cable life was observed for testing of new cable.
- New cable can be tested at 55 kV in the field prior to energization if aged cable has not been spliced in.
- Testing at lower dc voltages (such as 200 V/mil) will not pick out bad sections of cable.

Another option for testing cable integrity: ac testing does not degrade solid dielectric insulation (or at least degrades it more slowly). The use of very low frequency ac testing (at about 0.1 Hz) may cause less damage to aged cable than dc testing (Eager et al., 1997) (but utilities have reported that it is not totally benign, and ac testing has not gained widespread usage). The low frequency has the advantage that the equipment is much smaller than 60-Hz ac testing equipment.

3.9 Fault Location

Utilities use a variety of tools and techniques to locate underground faults. Several are described in the next few paragraphs [see also EPRI TR-105502 (1995)].

Divide and conquer — On a radial tap where the fuse has blown, crews narrow down the faulted section by opening the cable at locations. Crews start by opening the cable near the center, then they replace the fuse. If the fuse blows, the fault is upstream; if it doesn't blow, the fault is downstream. Crews then open the cable near the center of the remaining portion and continue bisecting the circuit at appropriate sectionalizing points (usually padmounted transformers). Of course, each time the cable faults, more damage is done at the fault location, and the rest of the system has the stress of carrying the fault currents. Using current-limiting fuses reduces the fault-current stress but increases the cost.

Fault indicators — Faulted circuit indicators (FCIs) are small devices clamped around a cable that measure current and signal the passage of fault current. Normally, these are applied at padmounted transformers. Faulted circuit indicators do not pinpoint the fault; they identify the fault to a cable section. After identifying the failed section, crews must use another method

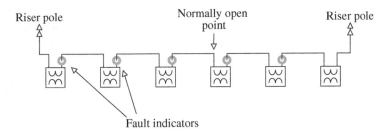

FIGURE 3.19
Typical URD fault indicator application.

such as the thumper to precisely identify the fault. If the entire section is in conduit, crews don't need to pinpoint the location; they can just pull the cable and replace it (or repair it if the faulted portion is visible from the outside). Cables in conduit require less precise fault location; a crew only needs to identify the fault to a given conduit section.

Utilities' main justification for faulted circuit indicators is reducing the length of customer interruptions. Faulted circuit indicators can significantly decrease the fault-finding stage relative to the divide-and-conquer method. Models that make an audible noise or have an external indicator decrease the time needed to open cabinets.

Utilities use most fault indicators on URD loops. With one fault indicator per transformer (see Figure 3.19), a crew can identify the failed section and immediately reconfigure the loop to restore power to all customers. The crew can then proceed to pinpoint the fault and repair it (or even delay the repair for a more convenient time). For larger residential subdivisions or for circuits through commercial areas, location is more complicated. In addition to transformers, fault indicators should be placed at each sectionalizing or junction box. On three-phase circuits, either a three-phase fault indicator or three single-phase indicators are available; single-phase indicators identify the faulted phase (a significant advantage). Other useful locations for fault indicators are on either end of cable sections of overhead circuits, which are common at river crossings or under major highways. These sections are not fused, but fault indicators will show patrolling crews whether the cable section has failed.

Fault indicators may be reset in a variety of ways. On manual reset units, crews must reset the devices once they trip. These units are less likely to reliably indicate faults. Self-resetting devices are more likely to be accurate as they automatically reset based on current, voltage, or time. Current-reset is most common; after tripping, if the unit senses current above a threshold, it resets [standard values are 3, 1.5, and 0.1 A (NRECA RER Project 90-8, 1993)]. With current reset, the minimum circuit load at that point must be above the threshold, or the unit will never reset. On URD loops, when applying current-reset indicators, consider that the open point might change. This changes the current that the fault indicator sees. Again, make sure the circuit load is enough to reset the fault indicator. Voltage reset models provide a voltage sensor; when the voltage exceeds some value (the voltage sensor senses at secondary voltage or at an elbow's capacitive test point). Time-reset units simply reset after a given length of time.

Fault indicators should only operate for faults — not for load, not for inrush, not for lightning, and not for backfeed currents. False readings can send crews on wild chases looking for faults. Reclose operations also cause loads and transformers to draw inrush, which can falsely trip a fault indicator. An inrush restraint feature disables tripping for up to one second following energization. On single-phase taps, inrush restraint is really only needed for manually-reset fault indicators (the faulted phase with the blown fuse will not have inrush that affects downstream fault indicators). Faults in adjacent cables can also falsely trip indicators; the magnetic fields couple into the pickup coil. Shielding can help prevent this. Several scenarios cause backfeed that can trip fault indicators. Downstream of a fault, the stored charge in the cable will rush into the fault, possibly tripping fault indicators. McNulty (1994) reported that 2000 ft of 15-kV cable created an oscillatory current transient that peaked at 100 A and decayed in 0.15 msec. Nearby capacitor banks on the overhead system can make outrush worse. Motors and other rotating equipment can also backfeed faults. To avoid false trips, use a high set point. Equipment with filtering that reduces the indicator's sensitivity to transient currents also helps, but too much filtering may leave the faulted-circuit indicator unable to detect faults cleared rapidly by current-limiting fuses.

Self-resetting fault indicators can also falsely reset. Backfeed currents and voltages can reset fault indicators. On a three-phase circuit with one phase tripped, the faulted phase can backfeed through three-phase transformer connections (see Chapter 4), providing enough current or enough voltage to reset faulted-circuit indicators. On single-phase circuits, these are not a problem. In general, single-phase application is much easier; we do not have backfeed problems or problems with indicators tripping from faults on nearby cables. For single-phase application guidelines, see (IEEE Std 1216-2000).

Fault indicators may have a threshold-type trip characteristic like an instantaneous relay (any current above the set point trips the flag), or they may have a time-overcurrent characteristic which trips faster for higher currents. Those units with time-overcurrent characteristics should be coordinated with minimum clearing curves of current-limiting fuses to ensure that they operate. Another type of fault indicator uses an adaptive setting that trips based on a sudden increase in current followed by a loss of current.

Set the trip level on fault indicators to less than 50% of the available fault current or 500 A, whichever is less (IEEE P1610/D03, 2002). This trip threshold should be at least two to three times the load on the circuit to minimize false indications. These two conditions will almost never conflict, only at the end of a very long feeder (low fault currents) on a cable that is heavily loaded.

Normally, fault indicators are fixed equipment, but they can be used for targeted fault location. When crews arrive at a faulted and isolated section, they first apply fault indicators between sections (normally at padmounted transformers). Crews reenergize the failed portion and then check the fault indicators to identify the faulted section. Only one extra fault is applied to the circuit, not multiple faults as with the divide and conquer method.

Section testing — Crews isolate a section of cable and apply a dc hi-pot voltage. If the cable holds the hi-pot voltage, crews proceed to the next section and repeat until finding a cable that cannot hold the hi-pot voltage. Because the voltage is dc, the cable must be isolated from the transformer. In a faster variation of this, high-voltage sticks are available that use the ac line voltage to apply a dc voltage to the isolated cable section.

Thumper — The thumper applies a pulsed dc voltage to the cable. As its name implies, at the fault the thumper discharges sound like a thumping noise as the gap at the failure point repeatedly sparks over. The thumper charges a capacitor and uses a triggered gap to discharge the capacitor's charge into the cable. Crews can find the fault by listening for the thumping noise. Acoustic enhancement devices can help crews locate weak thumping noises; antennas that pick up the radio-frequency interference from the arc discharge also help pinpoint the fault. Thumpers are good for finding the exact fault location so that crews can start digging. On a 15-kV class system, utilities typically thump with voltages from 10 to 15 kV, but utilities sometimes use voltages to 25 kV.

While pulsed discharges are thought to be less damaging to cable than a steady dc voltage, utilities have concern that thumping can damage the unfailed sections of cable. When a thumper pulse breaks down the cable, the incoming surge shoots past the fault. When it reaches the open point, the voltage doubles, then the voltage pulse bounces back and forth between the open point and the fault, switching from +2 to –2E (where E is the thumper pulse voltage). In tests, EPRI research found that thumping can reduce the life of aged cable (EPRI EL-6902, 1990; EPRI TR-108405-V1, 1997; Hartlein et al., 1989; Hartlein et al., 1994). The thumping discharges at the failure point can also increase the damage at the fault point. Most utilities try to limit the voltage or discharge energy, and a few don't use a thumper for fear of additional damage to cables and components (Tyner, 1998). A few utilities also disconnect transformers from the system during thumping to protect the transformer and prevent surges from propagating through the transformer (these surges should be small). If the fault has no gap, and if the fault is a solid short circuit, then no arc forms, and the thumper will not create its characteristic thump (fortunately, solid short circuits are rare in cable faults). Some crews keep thumping in an effort to burn the short circuit apart enough to start arcing. With cable in conduit, the thumping may be louder near the conduit ends than at the fault location. Generally, crews should start with the voltage low and increase as needed. A dc hi-pot voltage can help determine how much voltage the thumper needs.

Radar — Also called time-domain reflectometry (TDR), a radar set injects a very short-duration current pulse into the cable. At discontinuities, a portion of the pulse will reflect back to the set; knowing the velocity of wave propagation along cable gives us an estimate of the distance to the fault. Depending on the test set and settings, radar pulses can be from 5 ns to 5 μs wide. Narrower pulses give higher resolution, so users can better differentiate between faults and reflections from splices and other discontinuities (Banker et al., 1994).

Radar does not give pinpoint accuracy; its main use is to narrow the fault to a certain section. Then, crews can use a thumper or other pinpoint technique to find the failure. Taking a radar pulse from either end of a cable and averaging the results can lead to an improved estimate of the location. Radar location on circuits with taps can be complicated, especially those with multiple taps; the pulse will reflect off the taps, and the reflection from the actual fault will be less than it otherwise would be. Technology has been developed to use above-ground antennas to sense and pinpoint faults based on the radar signals.

Radar and thumper — After a fuse or other circuit interrupter clears a fault in a cable, the area around the fault point recovers some insulation strength. Checking the cable with an ohm meter would show an open circuit. Likewise, the radar pulse passes right by the fault, so the radar set alone cannot detect the fault. Using radar with a thumper solves this problem. A thumper pulse breaks down the gap, and the radar superimposes a pulse that reflects off the fault arc. The risetime of the thumper waveshape is on the order of a few microseconds; the radar pulse total width may be less than 0.05 μsec. Another less attractive approach is to use a thumper to continually burn the cable until the fault resistance becomes low enough to get a reading on a radar set (this is less attractive because it subjects the cable to many more thumps, especially if crews use high voltages).

Boucher (1991) reported that fault indicators were the most popular fault locating approach, but most utilities use a variety of techniques (see Figure 3.20). Depending on the type of circuit, the circuit layout, and the equipment available, different approaches are sometimes better.

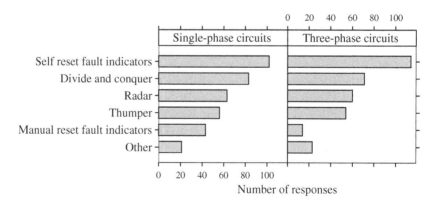

FIGURE 3.20
Utility use of fault-locating techniques (204 utilities surveyed, multiple responses allowed). (Data from [Boucher, 1991].)

When applying test voltages to cables, crews must be mindful that cables can hold significant charge. Cables have significant capacitance, and cables can maintain charge for days.

References

AEIC CS5-94, *Specification for Cross-Linked Polyethylene Insulated, Shielded Power Cables Rated 5 through 46 kV*, Association of Edison Illuminating Companies, 1994.

AEIC CS6-96, *Specification for Ethylene Propylene Rubber Shielded Power Cables Rated 5–69 kV*, Association of Edison Illuminating Companies, 1996.

Ametani, A., "A General Formulation of Impedance and Admittance of Cables," *IEEE Transactions on Power Apparatus and Systems*, vol. PAS-99, no. 3, pp. 902–10, May/June 1980.

Anders, G. and Brakelmann, H., "Cable Crossings-Derating Considerations. I. Derivation of Derating Equations," *IEEE Transactions on Power Delivery*, vol. 14, no. 3, pp. 709–14, July 1999a.

Anders, G. and Brakelmann, H., "Cable Crossings-Derating Considerations. II. Example of Derivation of Derating Curves," *IEEE Transactions on Power Delivery*, vol. 14, no. 3, pp. 715–20, July 1999b.

Anders, G. J., "Rating of Cables on Riser Poles, in Trays, in Tunnels and Shafts — A Review," *IEEE Transactions on Power Delivery*, vol. 11, no. 1, pp. 3–11, January 1996.

Anders, G. J., *Rating of Electric Power Cables: Ampacity Computations for Transmission, Distribution, and Industrial Applications*, IEEE Press, McGraw-Hill, New York, 1998.

ANSI/ICEA S-94-649-2000, *Standard for Concentric Neutral Cables Rated 5 through 46 kV*, Insulated Cable Engineers Association, 2000.

ANSI/ICEA S-97-682-2000, *Standard for Utility Shielded Power Cables Rated 5 through 46 kV*, Insulated Cable Engineers Association, 2000.

ANSI/IEEE Std. 400-1980, *IEEE Guide for Making High-Direct-Voltage Tests on Power Cable Systems in the Field*.

Banker, W. A., Nannery, P. R., Tarpey, J. W., Meyer, D. F., and Piesinger, G. H., "Application of High Resolution Radar to Provide Non-destructive Test Techniques for Locating URD Cable Faults and Splices," *IEEE Transactions on Power Delivery*, vol. 9, no. 3, pp. 1187–94, July 1994.

Boggs, S. and Xu, J. J., "Water Treeing — Filled vs. Unfilled Cable Insulation," *IEEE Electrical Insulation Magazine*, vol. 17, no. 1, pp. 23–9, January/February 2001.

Boucher, R., "A Summary Of The Regional Underground Distribution Practices For 1991," *Regional Underground Distribution Practices* (IEEE paper 91 TH0398-8-PWR), 1991.

Brown, M., "EPR Insulation Cuts Treeing and Cable Failures," *Electrical World*, vol. 197, no. 1, pp. 105–6, January 1983.

Brown, M., "Accelerated Life Testing of EPR-Insulated Underground Cable," *IEEE Electrical Insulation Magazine*, vol. 7, no. 4, pp. 21–6, July/August 1991.

Burns, N. M., Jr., "Performance of Supersmooth Extra Clean Semiconductive Shields in XLPE Insulated Power Cables," *IEEE International Symposium on Electrical Insulation*, 1990.

CEA 117 D 295, *Survey of Experience with Polymer Insulated Power Cable in Underground Service*, Canadian Electrical Association, 1987.

CEA 274 D 723, *Underground Versus Overhead Distribution Systems*, Canadian Electrical Association, 1992.

CEA, *CEA Distribution Planner's Manual*, Canadian Electrical Association, 1982.

Champion, T., "Elbow Failures Cast Doubt on Reliability," *Electrical World*, vol. 200, no. 6, pp. 71–3, June 1986.

Cinquemani, P. L., Yingli, W., Kuchta, F. L., and Doench, C., "Performance of Reduced Wall EPR Insulated Medium Voltage Power Cables. I. Electrical Characteristics," *IEEE Transactions on Power Delivery*, vol. 12, no. 2, pp. 571–8, April 1997.

Cress, S. L. and Motlis, H., "Temperature Rise of Submarine Cable on Riser Poles," *IEEE Transactions on Power Delivery*, vol. 6, no. 1, pp. 25–33, January 1991.

Dedman, J. C. and Bowles, H. L., "A Survey of URD Cable Installed on Rural Electric Systems and Failures of That Cable," IEEE Rural Electric Power Conference, 1990.

Dommel, H. W., "Electromagnetic Transients Program Reference Manual (EMTP Theory Book)," prepared for Bonneville Power Administration, 1986.

Dudas, J. H., "Technical Trends in Medium Voltage URD Cable Specifications," *IEEE Electrical Insulation Magazine*, vol. 10, no. 2, pp. 7–16, March/April 1994.

Dudas, J. H. and Cochran, W. H., "Technical Advances in the Underground Medium-Voltage Cable Specification of the Largest Investor-Owned Utilities in the U.S.," *IEEE Electrical Insulation Magazine*, vol. 15, no. 6, pp. 29–36, November/December 1999.

Dudas, J. H. and Rodgers, J. R., "Underground Cable Technical Trends for the Largest Rural Electric Co-ops," *IEEE Transactions on Industry Applications*, vol. 35, no. 2, pp. 324–31, March/April 1999.

Eager, G. S., Katz, C., Fryszczyn, B., Densley, J., and Bernstein, B. S., "High Voltage VLF Testing of Power Cables," *IEEE Transactions on Power Delivery*, vol. 12, no. 2, pp. 565–70, April 1997.

EPRI 1000273, *Estimation of Remaining Life of XLPE-Insulated Cables*, Electric Power Research Institute, Palo Alto, CA, 2000.

EPRI 1000741, *Condition Assessment of Distribution PILC Cable Assets*, Electric Power Research Institute, Palo Alto, CA, 2000.

EPRI 1001389, *Aging of Extruded Dielectric Distribution Cable: Phase 2 Service Aging*, Electric Power Research Institute, Palo Alto, CA, 2002.

EPRI 1001732, *Thermal Issues and Ratings of Separable Insulated Connectors*, Electric Power Research Institute, Palo Alto, CA, 2002.

EPRI 1001734, *State of the Art of Thin Wall Cables Including Industry Survey*, Electric Power Research Institute, Palo Alto, CA, 2002.

EPRI 1001894, *EPRI Power Cable Materials Selection Guide*, EPRI, Palo Alto, CA, 2001.

EPRI EL-6902, *Effects of Voltage Surges on Solid-Dielectric Cable Life*, Electric Power Research Institute, Palo Alto, CA, 1990.

EPRI TR-101245, *Effect of DC Testing on Extruded Cross-Linked Polyethylene Insulated Cables*, Electric Power Research Institute, Palo Alto, CA, 1993.

EPRI TR-101245-V2, *Effect of DC Testing on Extruded Cross-Linked Polyethylene Insulated Cables — Phase II*, Electric Power Research Institute, Palo Alto, CA, 1995.

EPRI TR-105502, *Underground Cable Fault Location Reference Manual*, Electric Power Research Institute, Palo Alto, CA, 1995.

EPRI TR-108405-V1, *Aging Study of Distribution Cables at Ambient Temperatures with Surges*, Electric Power Research Institute, Palo Alto, CA, 1997.

EPRI TR-108919, *Soil Thermal Properties Manual for Underground Power Transmission,* Electric Power Research Institute, Palo Alto, CA, 1997.

EPRI TR-111888, *High-Ampacity, Thin-Wall, Novel Polymer Cable,* Electric Power Research Institute, Palo Alto, CA, 2000.

Gurniak, B., "Neutral Corrosion Problem Overstated," *Transmission & Distribution World,* pp. 152–8, August 1996.

Hartlein, R. A. and Black, W. Z., "Ampacity of Electric Power Cables in Vertical Protective Risers," *IEEE Transactions on Power Apparatus and Systems,* vol. PAS-102, no. 6, pp. 1678–86, June 1983.

Hartlein, R. A., Harper, V. S., and Ng, H. W., "Effects of Voltage Impulses on Extruded Dielectric Cable Life," *IEEE Transactions on Power Delivery,* vol. 4, no. 2, pp. 829–41, April 1989.

Hartlein, R. A., Harper, V. S., and Ng, H. W., "Effects of Voltage Surges on Extruded Dielectric Cable Life Project Update," *IEEE Transactions on Power Delivery,* vol. 9, no. 2, pp. 611–9, April 1994.

Horton, W. F. and Golberg, S., "The Failure Rates of Underground Distribution System Components," Proceedings of the Twenty-Second Annual North American Power Symposium, 1990.

Horton, W. F. and Golberg, S., "Determination of Failure Rates of Underground Distribution System Components from Historical Data," IEEE/PES Transmission and Distribution Conference, 1991.

ICEA Publication T-31-610, *Guide for Conducting a Longitudinal Water Penetration Resistance Test for Sealed Conductor,* Insulated Cable Engineers Association, 1994.

ICEA Publication T-34-664, *Conducting Longitudinal Water Penetration Resistance Tests on Cable,* Insulated Cable Engineers Association, 1996.

ICEA S-66-524, *Cross-Linked Thermosetting Polyethylene Insulated Wire and Cable for Transmission and Distribution of Electrical Energy,* Insulated Cable Engineers Association, 1988.

IEC 287, *Calculation of the Continuous Current Rating of Cables (100% Load Factor),* 2nd ed., International Electrical Commission (IEC), 1982.

IEEE C2-1997, National Electrical Safety Code.

IEEE P1610/D03, *Draft Guide for the Application of Faulted Circuit Indicators for 200/600 A, Three-Phase Underground Distribution,* 2002.

IEEE Std 1216-2000, *IEEE Guide for the Application of Faulted Circuit Indicators for 200 A, Single-Phase Underground Residential Distribution (URD).*

IEEE Std. 141-1993, *IEEE Recommended Practice for Electric Power Distribution for Industrial Plants.*

IEEE Std. 386-1995, *IEEE Standard for Separable Insulated Connector Systems for Power Distribution Systems Above 600 V.*

IEEE Std. 835-1994, *IEEE Standard Power Cable Ampacities.*

Katz, C. and Walker, M., "Evaluation of Service Aged 35 kV TR-XLPE URD Cables," *IEEE Transactions on Power Delivery,* vol. 13, no. 1, pp. 1–6, January 1998.

Kerite Company, "Technical Application Support Data." Downloaded from www.kerite.com, June 2002.

Koch, B., "Underground Lines: Different Problems, Practical Solutions," *Electrical World T&D,* March/April 2001.

Lewis, W. A. and Allen, G. D., "Symmetrical Component Circuit Constants and Neutral Circulating Currents for Concentric Neutral Underground Distribution Cables," *IEEE Transactions on Power Apparatus and Systems,* vol. PAS-97, no. 1, pp. 191–9, January/February 1978.

Lewis, W. A., Allen, G. D., and Wang, J. C., "Circuit Constants for Concentric Neutral Underground Distribution Cables on a Phase Basis," *IEEE Transactions on Power Apparatus and Systems*, vol. PAS-97, no. 1, pp. 200–7, January/February 1978.

Mackevich, J. P., "Trends in Underground Residential Cable Systems," IEEE Rural Electric Power Conference, 1988.

Martin, M. A., Silver, D. A., Lukac, R. G., and Suarez, R., "Normal and Short Circuit Operating Characteristics of Metallic Shielding Solid Dielectric Power Cable," *IEEE Transactions on Power Apparatus and Systems*, vol. PAS-93, no. 2, pp. 601–13, March/April 1974.

McNulty, W. J., "False Tripping of Faulted Circuit Indicators," IEEE/PES Transmission and Distribution Conference, 1994.

Mercier, C. D. and Ticker, S., "DC Field Test for Medium-Voltage Cables: Why Can No One Agree?," *IEEE Transactions on Industry Applications*, vol. 34, no. 6, pp. 1366–70, November/December 1998.

Nannery, P. R., Tarpey, J. W., Lacenere, J. S., Meyer, D. F., and Bertini, G., "Extending the Service Life of 15 kV Polyethylene URD Cable Using Silicone Liquid," *IEEE Transactions on Power Delivery*, vol. 4, no. 4, pp. 1991–6, October 1989.

Neher, J. H. and McGrath, M. H., "The Calculation of the Temperature Rise and Load Capability of Cable Systems," *AIEE Transactions*, vol. 76, pp. 752–64, October 1957.

NFPA 70, *National Electrical Code*, National Fire Protection Association, 1999.

NRECA RER Project 90-8, *Underground Distribution System Design and Installation Guide*, National Rural Electric Cooperative Association, 1993.

Okonite, *Engineering Data for Copper and Aluminum Conductor Electrical Tables*, Okonite Company, publication EHB-90, 1990.

Powers, W. F., "An Overview of Water-Resistant Cable Designs," *IEEE Transactions on Industry Applications*, vol. 29, no. 5, 1993.

Smith, D. R., "System Considerations — Impedance and Fault Current Calculations," IEEE Tutorial Course on Application and Coordination of Reclosers, Sectionalizers, and Fuses, 1980. Publication 80 EHO157-8-PWR.

Smith, D. R. and Barger, J. V., "Impedance and Circulating Current Calculations for UD Multi-Wire Concentric Neutral Circuits," *IEEE Transactions on Power Apparatus and Systems*, vol. PAS-91, no. 3, pp. 992–1006, May/June 1972.

Southwire Company, *Power Cable Manual*, 2nd ed, 1997.

St. Pierre, C., *A Practical Guide to Short-Circuit Calculations*, Electric Power Consultants, Schenectady, NY, 2001.

State of New York Department of Public Service, "A Report on Consolidated Edison's July 1999 System Outages," March 2000.

Stember, L. H., Epstein, M. M., Gaines, G. V., Derringer, G. C., and Thomas, R. E., "Analysis of Field Failure Data on HMWPE- and XLPE-Insulated High-Voltage Distribution Cable," *IEEE Transactions on Power Apparatus and Systems*, vol. PAS-104, no. 8, pp. 1979–85, August 1985.

Stringer, N. T. and Kojovic, L. A., "Prevention of Underground Cable Splice Failures," *IEEE Transactions on Industry Applications*, vol. 37, no. 1, pp. 230–9, January/February 2001.

Thue, W. A., *Electrical Power Cable Engineering*, Marcel Dekker, New York, 1999.

Tyner, J. T., "Getting to the Bottom of UG Practices," *Transmission & Distribution World*, vol. 50, no. 7, pp. 44–56, July 1998.

Westinghouse Electric Corporation, *Electrical Transmission and Distribution Reference Book*, 1950.

I was down in the hole and pulled on one of the splices thinking that I might find the faulted one by pulling it apart, well you can only guess what happens next! KA-BOOOM, I was really pissed off at that point and still am. BUT YOU KNOW SOMETHING, IT'S MY FAULT FOR TAKING SOME OTHER HALF ASS LINEMAN'S WORD FOR IT BEING DEAD AND NOT CHECKING IT OUT FOR MY SELF!

Anonymous poster, about beginning work after another lineman told him that the cables were disconnected at the source end.

www.powerlineman.com

4

Transformers

Ac transformers are one of the keys to allowing widespread distribution of electric power as we see it today. Transformers efficiently convert electricity to higher voltage for long distance transmission and back down to low voltages suitable for customer usage. The distribution transformer normally serves as the final transition to the customer and often provides a local grounding reference. Most distribution circuits have hundreds of distribution transformers. Distribution feeders may also have other transformers: voltage regulators, feeder step banks to interface circuits of different voltages, and grounding banks.

4.1 Basics

A transformer efficiently converts electric power from one voltage level to another. A transformer is two sets of coils coupled together through a magnetic field. The magnetic field transfers all of the energy (except in an autotransformer). In an ideal transformer, the voltages on the input and the output are related by the turns ratio of the transformer:

$$V_1 = \frac{N_1}{N_2} V_2$$

where N_1 and N_2 are the number of turns and V_1 and V_2 are the voltage on windings 1 and 2.

In a real transformer, not all of the flux couples between windings. This *leakage* flux creates a voltage drop between windings, so the voltage is more accurately described by

$$V_1 = \frac{N_1}{N_2} V_2 - X_L I_1$$

where X_L is the leakage reactance in ohms as seen from winding 1, and I_1 is the current out of winding 1.

The current also transforms by the turns ratio, opposite of the voltage as

$$I_1 = \frac{N_2}{N_1} I_2 \quad \text{or} \quad N_1 I_1 = N_2 I_2$$

The "ampere-turns" stay constant at $N_1 I_1 = N_2 I_2$; this fundamental relationship holds well for power and distribution transformers.

A transformer has a magnetic core that can carry large magnetic fields. The cold-rolled, grain-oriented steels used in cores have permeabilities of over 1000 times that of air. The steel provides a very low-reluctance path for magnetic fields created by current through the windings.

Consider voltage applied to the *primary* side (source side, high-voltage side) with no load on the *secondary* side (load side, low-voltage side). The winding draws *exciting* current from the system that sets up a sinusoidal magnetic field in the core. The flux in turn creates a back emf in the coil that limits the current drawn into the transformer. A transformer with no load on the secondary draws very little current, just the exciting current, which is normally less than 0.5% of the transformer's full-load current. On the unloaded secondary, the sinusoidal flux creates an open-circuit voltage equal to the primary-side voltage times the turns ratio.

When we add load to the secondary of the transformer, the load pulls current through the secondary winding. The magnetic coupling of the secondary current pulls current through the primary winding, keeping constant ampere-turns. Normally in an inductive circuit, higher current creates more flux, but not in a transformer (except for the leakage flux). The increasing force from current in one winding is countered by the decreasing force from current through the other winding (see Figure 4.1). The flux in the core on a loaded transformer is the same as that on an unloaded transformer, even though the current is much higher.

The voltage on the primary winding determines the flux in the transformer (the flux is proportional to the time integral of voltage). The flux in the core determines the voltage on the output-side of the transformer (the voltage is proportional to the time derivative of the flux).

Figure 4.2 shows models with the significant impedances in a transformer. The detailed model shows the series impedances, the resistances and the reactances. The series resistance is mainly the resistance of the wires in each winding. The series reactance is the leakage impedance. The shunt branch is the magnetizing branch, current that flows to magnetize the core. Most of the magnetizing current is reactive power, but it includes a real power component. Power is lost in the core through:

- *Hysteresis* — As the magnetic dipoles change direction, the core heats up from the friction of the molecules.

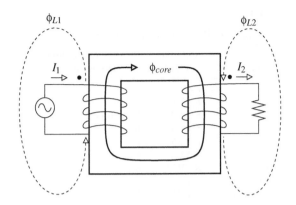

Magnetic equivalent circuit

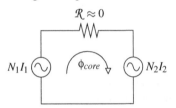

Since $\mathcal{R} \approx 0$, $N_1 I_1 = N_2 I_2$

Electric circuit

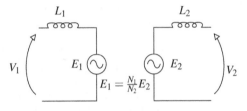

L_1 and L_2 are from the leakage fluxes, ϕ_{L1} and ϕ_{L2}

FIGURE 4.1
Transformer basic function.

Detailed transformer model

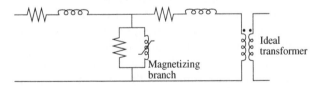

Simplified model

FIGURE 4.2
Transformer models.

TABLE 4.1

Common Scaling Ratios in Transformers

Quantity	Relative to kVA	Relative to a Reference Dimension, l
Rating	kVA	l^4
Weight	K kVA$^{3/4}$	K l^3
Cost	K KVA$^{3/4}$	K (% Total Loss)$^{-3}$
Length	K kVA$^{1/4}$	K l
Width	K kVA$^{1/4}$	K l
Height	K kVA$^{1/4}$	K l
Total losses	K kVA$^{3/4}$	K l^3
No-load losses	K kVA$^{3/4}$	K l^3
Exciting current	K kVA$^{3/4}$	K l^3
% Total loss	K kVA$^{-1/4}$	K l^{-1}
% No-load loss	K kVA$^{-1/4}$	K l^{-1}
% Exciting current	K kVA$^{-1/4}$	K l^{-1}
% R	K kVA$^{-1/4}$	K l^{-1}
% X	K kVA$^{1/4}$	K l
Volts/turn	K kVA$^{1/2}$	K l^2

Source: Arthur D. Little, "Distribution Transformer Rulemaking Engineering Analysis Update," Report to U.S. Department of Energy Office of Building Technology, State, and Community Programs. Draft. December 17, 2001.

- *Eddy currents* — Eddy currents in the core material cause resistive losses. The core flux induces the eddy currents tending to oppose the change in flux density.

The magnetizing branch impedance is normally above 5,000% on a transformer's base, so we can neglect it in many cases. The core losses are often referred to as iron losses or no-load losses. The load losses are frequently called the wire losses or copper losses. The various parameters of transformers scale with size differently as summarized in Table 4.1.

The simplified transformer model in Figure 4.2 with series resistance and reactance is sufficient for most calculations including load flows, short-circuit calculations, motor starting, or unbalance. Small distribution transformers have low leakage reactances, some less than 1% on the transformer rating, and X/R ratios of 0.5 to 5. Larger power transformers used in distribution substations have higher impedances, usually on the order of 7 to 10% with X/R ratios between 10 and 40.

The leakage reactance causes voltage drop on a loaded transformer. The voltage is from flux that doesn't couple from the primary to the secondary winding. Blume et al. (1951) describes leakage reactance well. In a real transformer, the windings are wound around a core; the high- and low-voltage windings are adjacent to each other. Figure 4.3 shows a configuration; each winding contains a number of turns of wire. The sum of the current in each wire of the high-voltage winding equals the sum of the currents in the

Side View of Windings   Top View of Windings

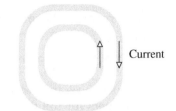

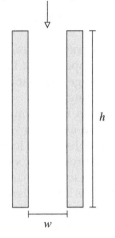

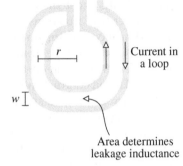

FIGURE 4.3
Leakage reactance.

low-voltage winding ($N_1 I_1 = N_2 I_2$), so each winding is equivalent to a busbar. Each busbar carries equal current, but in opposite directions. The opposing currents create flux in the gap between the windings (this is called *leakage flux*). Now, looking at the two windings from the top, we see that the windings are equivalent to current flowing in a loop encompassing a given area. This area determines the leakage inductance.

The leakage reactance in percent is based on the coil parameters and separations (Blume et al., 1951) as follows:

$$X_\% = \frac{126f(NI)^2 rw}{10^{11} h S_{kVA}}$$

where
f = system frequency, Hz
N = number of turns on one winding
I = full load current on the winding, A
r = radius to the windings, in.
w = width between windings, in.
h = height of the windings, in.
S_{kVA} = transformer rating, kVA

In general, leakage impedance increases with:

- Higher primary voltage (thicker insulation between windings)
- kVA rating
- Larger core (larger diameter leads to more area enclosed)

Leakage impedances are under control of the designer, and companies will make transformers for utilities with customized impedances. Large distribution substation transformers often need high leakage impedance to control fault currents, some as high as 30% on the base rating.

Mineral oil fills most distribution and substation transformers. The oil provides two critical functions: conducting heat and insulation. Because the oil is a good heat conductor, an oil-filled transformer has more load-carrying capability than a dry-type transformer. Since it provides good electrical insulation, clearances in an oil-filled transformer are smaller than a dry-type transformer. The oil conducts heat away from the coils into the larger thermal mass of the surrounding oil and to the transformer tank to be dissipated into the surrounding environment. Oil can operate continuously at high temperatures, with a normal operating temperature of 105°C. It is flammable; the flash point is 150°C, and the fire point is 180°C. Oil has high dielectric strength, 220 kV/in. (86.6 kV/cm), and evens out voltage stresses since the dielectric constant of oil is about 2.2, which is close to that of the insulation. The oil also coats and protects the coils and cores and other metal surfaces from corrosion.

4.2   Distribution Transformers

From a few kVA to a few MVA, distribution transformers convert primary-voltage to low voltage that customers can use. In North America, 40 million distribution transformers are in service, and another one million are installed each year (Alexander Publications, 2001). The transformer connection determines the customer's voltages and grounding configuration.

Distribution transformers are available in several standardized sizes as shown in Table 4.2. Most installations are single phase. The most common

TABLE 4.2

Standard Distribution Transformer Sizes

Distribution Transformer Standard Ratings, kVA	
Single phase	5, 10, 15, 25, 37.5, 50, 75, 100, 167, 250, 333, 500
Three phase	30, 45, 75, 112.5, 150, 225, 300, 500

TABLE 4.3

Insulation Levels for Distribution Transformers

Low-Frequency Test Level, kV rms	Basic Lightning Impulse Insulation Level, kV Crest	Chopped-Wave Impulse Levels	
		Minimum Voltage, kV Crest	Minimum Time to Flashover, µs
10	30	36	1.0
15	45	54	1.5
19	60	69	1.5
26	75	88	1.6
34	95	110	1.8
40	125	145	2.25
50	150	175	3.0
70	200	230	3.0
95	250	290	3.0
140	350	400	3.0

Source: IEEE Std. C57.12.00-2000. Copyright 2000 IEEE. All rights reserved.

overhead transformer is the 25-kVA unit; padmounted transformers tend to be slightly larger where the 50-kVA unit is the most common.

Distribution transformer impedances are rather low. Units under 50 kVA have impedances less than 2%. Three-phase underground transformers in the range of 750 to 2500 kVA normally have a 5.75% impedance as specified in (ANSI/IEEE C57.12.24-1988). Lower impedance transformers provide better voltage regulation and less voltage flicker for motor starting or other fluctuating loads. But lower impedance transformers increase fault currents on the secondary, and secondary faults impact the primary side more (deeper voltage sags and more fault current on the primary).

Standards specify the insulation capabilities of distribution transformer windings (see Table 4.3). The low-frequency test is a power-frequency (60 Hz) test applied for one minute. The basic lightning impulse insulation level (BIL) is a fast impulse transient. The front-of-wave impulse levels are even shorter-duration impulses.

The through-fault capability of distribution transformers is also given in IEEE C57.12.00-2000 (see Table 4.4). The duration in seconds of the short-circuit capability is:

$$t = \frac{1250}{I^2}$$

where I is the symmetrical current in multiples of the normal base current from Table 4.4.

Overhead and padmounted transformer tanks are normally made of mild carbon steel. Corrosion is one of the main concerns, especially for anything on the ground or in the ground. Padmounted transformers tend to corrode

TABLE 4.4

Through-Fault Capability of Distribution Transformers

Single-Phase Rating, kVA	Three-Phase Rating, kVA	Withstand Capability in per Unit of Base Current (Symmetrical)
5–25	15–75	40
37.5–110	112.5–300	35
167–500	500	25

Source: IEEE Std. C57.12.00-2000, *IEEE Standard General Requirements for Liquid-Immersed Distribution, Power, and Regulating Transformers.*

near the base (where moisture and dirt and other debris may collect). Submersible units, being highly susceptible to corrosion, are often stainless steel.

Distribution transformers are "self cooled"; they do not have extra cooling capability like power transformers. They only have one kVA rating. Because they are small and because customer peak loadings are relatively short duration, overhead and padmounted distribution transformers have significant overload capability. Utilities regularly size them to have peak loads exceeding 150% of the nameplate rating.

Transformers in underground vaults are often used in cities, especially for network transformers (feeding secondary grid networks). In this application, heat can be effectively dissipated (but not as well as with an overhead or padmounted transformer).

Subsurface transformers are installed in an enclosure just big enough to house the transformer with a grate covering the top. A "submersible" transformer is normally used, one which can be submerged in water for an extended period (ANSI/IEEE C57.12.80-1978). Heat is dissipated through the grate at the top. Dirt and debris in the enclosure can accelerate corrosion. Debris blocking the grates or vents can overheat the transformer.

Direct-buried transformers have been attempted over the years. The main problems have been overheating and corrosion. In soils with high electrical and thermal resistivity, overheating is the main concern. In soils with low electrical and thermal resistivity, overheating is not as much of a concern, but corrosion becomes a problem. Thermal conductivity in a direct-buried transformer depends on the thermal conductivity of the soil. The buried transformer generates enough heat to dry out the surrounding soil; the dried soil shrinks and creates air gaps. These air gaps act as insulating layers that further trap heat in the transformer.

4.3   Single-Phase Transformers

Single-phase transformers supply single-phase service; we can use two or three single-phase units in a variety of configurations to supply three-phase

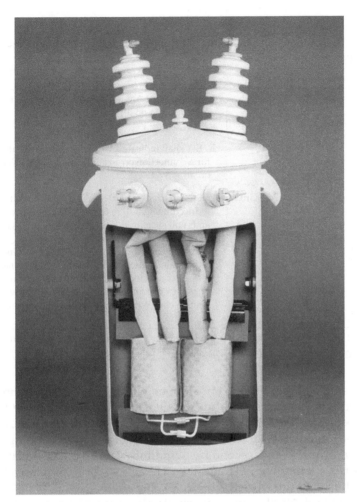

FIGURE 4.4
Single-phase distribution transformer. (Photo courtesy of ABB, Inc. With permission.)

service. A transformer's nameplate gives the kVA ratings, the voltage ratings, percent impedance, polarity, weight, connection diagram, and cooling class. Figure 4.4 shows a cutaway view of a single-phase transformer.

For a single-phase transformer supplying single-phase service, the load-full current in amperes is

$$I = \frac{S_{kVA}}{V_{kV}}$$

where
S_{kVA} = Transformer kVA rating
V_{kV} = Line-to-ground voltage rating in kV

TABLE 4.5

Winding Designations for Single-Phase Primary and Secondary Transformer Windings with One Winding

Nomenclature	Examples	Description
E	13800	E shall indicate a winding of E volts that is suitable for Δ connection on an E volt system.
E/E_1Y	2400/4160Y	E/E_1Y shall indicate a winding of E volts that is suitable for Δ connection on an E volt system or for Y connection on an E_1 volt system.
E/E_1GrdY	7200/12470GrdY	E/E_1GrdY shall indicate a winding of E volts having reduced insulation that is suitable for Δ connection on an E volt system or Y connection on an E_1 volt system, transformer, neutral effectively grounded.
E_1GrdY/E	12470GrdY/7200 480GrdY/277	E_1GrdY/E shall indicate a winding of E volts with reduced insulation at the neutral end. The neutral end may be connected directly to the tank for Y or for single-phase operation on an E_1 volt system, provided the neutral end of the winding is effectively grounded.
$E_1 = \sqrt{3}\ E$		

Note: E is line-to-neutral voltage of a Y winding, or line-to-line voltage of a Δ winding.

Source: IEEE Std. C57.12.00-2000. Copyright 2000 IEEE. All rights reserved.

So, a single-phase 50-kVA transformer with a high-voltage winding of 12470GrdY/7200 V has a full-load current of 6.94 A on the primary. On a 240/120-V secondary, the full-load current across the 240-V winding is 208.3 A.

Table 4.5 and Table 4.6 show the standard single-phase winding connections for primary and secondary windings. High-voltage bushings are labeled H*, starting with H1 and then H2 and so forth. Similarly, the low-voltage bushings are labeled X1, X2, X3, and so on.

The standard North American single-phase transformer connection is shown in Figure 4.5. The standard secondary load service is a 120/240-V three-wire service. This configuration has two secondary windings in series with the midpoint grounded. The secondary terminals are labeled X1, X2, and X3 where the voltage X1-X2 and X2-X3 are each 120 V. X1-X3 is 240 V.

Power and distribution transformers are assigned polarity dots according to the terminal markings. Current entering H1 results in current leaving X1. The voltage from H1 to H2 is in phase with the voltage from X1 to X3.

On overhead distribution transformers, the high-voltage terminal H1 is always on the left (when looking into the low-voltage terminals; the terminals are not marked). On the low-voltage side, the terminal locations are different, depending on size. If X1 is on the right, it is referred to as *additive polarity* (if X3 is on the right, it is *subtractive polarity*). Polarity is additive if the voltages add when the two windings are connected in series around the transformer (see Figure 4.6). Industry standards specify the polarity of a

TABLE 4.6

Two-Winding Transformer Designations for Single-Phase Primaries and Secondaries

Nomenclature	Examples	Description
E/2E	120/240 240/280 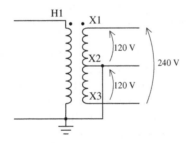 X4 X3 X2 X1	E/2E shall indicate a winding, the sections of which can be connected in parallel for operation at E volts, or which can be connected in series for operation at 2E volts, or connected in series with a center terminal for three-wire operation at 2E volts between the extreme terminals and E volts between the center terminal and each of the extreme terminals.
2E/E	240/120 X3 X2 X1	2E/E shall indicate a winding for 2E volts, two-wire full kilovoltamperes between extreme terminals, or for 2E/E volts three-wire service with 1/2 kVA available only, from midpoint to each extreme terminal.
E × 2E	240 × 480 X4 X3 X2 X1	E × 2E shall indicate a winding for parallel or series operation only but not suitable for three-wire service.

Source: IEEE Std. C57.12.00-2000. Copyright 2000 IEEE. All rights reserved.

FIGURE 4.5
Single-phase distribution transformer diagram.

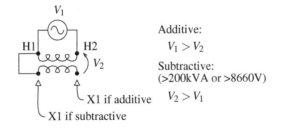

Additive:
$V_1 > V_2$

Subtractive:
(>200kVA or >8660V)
$V_2 > V_1$

FIGURE 4.6
Additive and subtractive polarity.

transformer, which depends on the size and the high-voltage winding. Single-phase transformers have additive polarity if (IEEE C57.12.00-2000):

$$kVA \leq 200 \text{ and } V \leq 8660$$

All other distribution transformers have subtractive polarity. The reason for the division is that originally all distribution transformers had additive polarity and all power transformers had subtractive polarity. Increasing sizes of distribution transformers caused overlap between "distribution" and "power" transformers, so larger distribution transformers were made with subtractive polarity for consistency. Polarity is important when connecting single-phase units in three-phase banks and for paralleling units.

Manufacturers make single-phase transformers as either shell form or core form (see Figure 4.7). Core-form designs prevailed prior to the 1960s; now, both shell- and core-form designs are available. Single-phase core-form transformers must have *interlaced* secondary windings (the low-high-low design). Every secondary leg has two coils, one wrapped around each leg of the core. The balanced configuration of the interlaced design allows unbalanced loadings on each secondary leg. Without interlacing, unbalanced secondary loads excessively heat the tank. An unbalanced secondary load creates an unbalanced flux in the iron core. The core-form construction does not have a return path for the unbalanced flux, so the flux returns outside of the iron core (in contrast, the shell-form construction has a return path for such flux). Some of the stray flux loops through the transformer tank and heats the tank.

The shell-form design does not need to have interlaced windings, so the *noninterlaced* configuration is normally used on shell-form transformers since it is simpler. The noninterlaced secondary has two to four times the reactance: the secondary windings are separated by the high-voltage winding and the insulation between them. Interlacing reduces the reactance since the low-voltage windings are right next to each other.

Using a transformer's impedance magnitude and load losses, we can find the real and reactive impedance in percent as

$$R = \frac{W_{CU}}{10S_{kVA}}$$

$$X = \sqrt{Z^2 - R^2}$$

where
 S_{kVA} = transformer rating, kVA
 $W_{CU} = W_{TOT} - W_{NL}$ = load loss at rated load, W
 W_{TOT} = total losses at rated load, W
 W_{NL} = no-load losses, W
 Z = nameplate impedance magnitude, %

Core form, interlaced

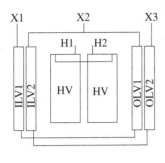
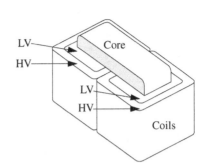

Shell form, non-interlaced

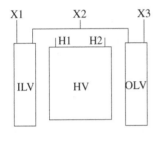
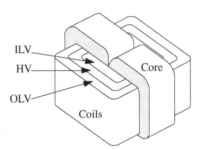

FIGURE 4.7
Core-form and shell-form single-phase distribution transformers. (From IEEE Task Force Report, "Secondary (Low-Side) Surges in Distribution Transformers," *IEEE Trans. Power Delivery,* 7(2), 746–756, April 1992. With permission. ©1992 IEEE.)

The nameplate impedance of a single-phase transformer is the *full-winding* impedance, the impedance seen from the primary when the full secondary winding is shorted from X1 to X3. Other impedances are also important; we need the two *half-winding* impedances for secondary short-circuit calculations and for unbalance calculations on the secondary. One impedance is the impedance seen from the primary for a short circuit from X1 to X2. Another is from X2 to X3. The half-winding impedances are not provided on the nameplate; we can measure them or use the following approximations. Figure 4.8 shows a model of a secondary winding for use in calculations.

The half-winding impedance of a transformer depends on the construction. In the model in Figure 4.8, one of the half-winding impedances in percent equals $Z_A + Z_1$; the other equals $Z_A + Z_2$. A core- or shell-form transformer with an interlaced secondary winding has an impedance in percent of approximately:

$$Z_{HX1-2} = Z_{HX2-3} = 1.5\,R + j\,1.2\,X$$

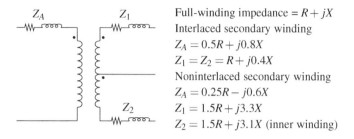

$$\text{Full-winding impedance} = R + jX$$
Interlaced secondary winding
$$Z_A = 0.5R + j0.8X$$
$$Z_1 = Z_2 = R + j0.4X$$
Noninterlaced secondary winding
$$Z_A = 0.25R - j0.6X$$
$$Z_1 = 1.5R + j3.3X$$
$$Z_2 = 1.5R + j3.1X \text{ (inner winding)}$$

FIGURE 4.8
Model of a 120/240-V secondary winding with all impedances in percent. (Impedance data from [Hopkinson, 1976].)

where R and X are the real and reactive components of the full-winding impedance (H1 to H2 and X1 to X3) in percent. A noninterlaced shell-form transformer has an impedance in percent of approximately:

$$Z_{HX1-2} = Z_{HX2-3} = 1.75\ R + j\ 2.5\ X$$

In a noninterlaced transformer, the two half-winding impedances are not identical; the impedance to the inner low-voltage winding is less than the impedance to the outer winding (the radius to the gap between the outer secondary winding and the primary winding is larger, so the gap between windings has more area).

A secondary fault across one 120-V winding at the terminals of a noninterlaced transformer has current about equal to the current for a fault across the 240-V winding. On an interlaced transformer, the lower relative impedance causes higher currents for the 120-V fault.

Consider a 50-kVA transformer with Z = 2%, 655 W of total losses, no-load losses of 106 W, and a noninterlaced 120/240-V secondary winding. This translates into a full-winding percent impedance of 1.1 + j1.67. For a fault across the 240-V winding, the current is found as

$$Z_{\Omega,240V} = (R + jX)\frac{10(0.24kV)^2}{S_{kVA}} = (1.1 + j1.67)\frac{10(0.24kV)^2}{50kVA} = 0.013 + j0.019\Omega$$

$$I_{240V} = \left|\frac{0.24kV}{Z_{\Omega,240V}}\right| = 10.4kA$$

For a fault across the 120-V winding on this noninterlaced transformer, the current is found as

$$Z_{\Omega,120V} = (1.75R + j2.5X)\frac{10(0.12\text{kV})^2}{S_{kVA}} = (1.93 + j4.18)\frac{10(0.12\text{kV})^2}{50\text{kVA}}$$

$$= 0.0055 + j0.0120\,\Omega$$

$$I_{120V} = \left|\frac{0.12\text{kV}}{Z_{\Omega,120V}}\right| = 9.06\text{kA}$$

Consider the same transformer characteristics on a transformer with an interlaced secondary and $Z = 1.4\%$. The 240-V and 120-V short-circuit currents are found as

$$Z_{\Omega,240V} = (R + jX)\frac{10(0.24\text{kV})^2}{S_{kVA}} = (1.1 + j0.87)\frac{10(0.24\text{kV})^2}{50\text{kVA}} = 0.013 + j0.01\,\Omega$$

$$I_{240V} = \left|\frac{0.24\text{kV}}{Z_{\Omega,240V}}\right| = 14.9\text{kA}$$

$$Z_{\Omega,120V} = (1.5R + j1.2X)\frac{10(0.12\text{kV})^2}{S_{kVA}} = (1.65 + j1.04)\frac{10(0.12\text{kV})^2}{50\text{kVA}}$$

$$= 0.0048 + j0.003\,\Omega$$

$$I_{120V} = \left|\frac{0.12\text{kV}}{Z_{\Omega,120V}}\right| = 21.4\text{kA}$$

The fault current for a 120-V fault is significantly higher than the 240-V current.

Completely self-protected transformers (CSPs) are a widely used single-phase distribution transformer with several built-in features (see Figure 4.9):

- Tank-mounted arrester
- Internal "weak-link" fuse
- Secondary breaker

CSPs do not need a primary-side cutout with a fuse. The internal primary fuse protects against an internal failure in the transformer. The weak link has less fault-clearing capability than a fuse in a cutout, so they need external current-limiting fuses where fault currents are high.

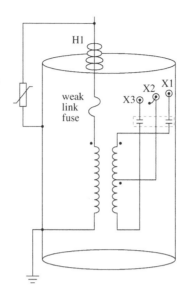

FIGURE 4.9
Completely self-protected transformer.

Secondary breakers provide protection against overloads and secondary faults. The breaker responds to current and oil temperature. Tripping is controlled by deflection of bimetallic elements in series. The oil temperature and current through the bimetallic strips heat the bimetal. Past a critical temperature, the bimetallic strips deflect enough to operate the breaker. Figure 4.10 shows trip characteristics for secondary breakers inside two size transformers. The secondary breaker has an emergency position to allow extra overload without tripping (to allow crews time to replace the unit). Crews can also use the breaker to drop the secondary load.

Some CSPs have overload-indicating lights that signal an overload. The indicator light doesn't go off until line crews reset the breaker. The indicator lights are not ordered as often (and crews often disable them in the field) because they generate a fair number of nuisance phone calls from curious/helpful customers.

4.4 Three-Phase Transformers

Three-phase overhead transformer services are normally constructed from three single-phase units. Three-phase transformers for underground service (either padmounted, direct buried, or in a vault or building or manhole) are normally single units, usually on a three- or five-legged core. Three-phase distribution transformers are usually core construction (see Figure 4.11), with

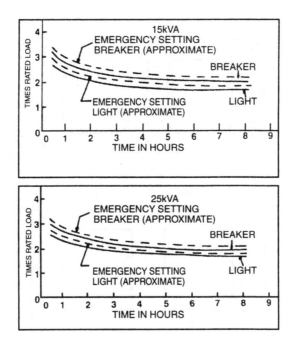

FIGURE 4.10
Clearing characteristics of a secondary breaker. (From ERMCO, Inc. With permission.)

either a three-, four-, or five-legged core construction (shell-type construction is rarely used). The five-legged wound core transformer is very common. Another option is *triplex* construction, where the three transformer legs are made from single individual core/coil assemblies (just like having three separate transformers).

The kVA rating for a three-phase bank is the total of all three phases. The full-load current in amps in each phase of a three-phase unit or bank is

$$I = \frac{S_{kVA}}{3V_{LG,kV}} = \frac{S_{kVA}}{\sqrt{3}V_{LL,kV}}$$

where
S_{kVA} = Transformer three-phase kVA rating
$V_{LG,kV}$ = Line-to-ground voltage rating, kV
$V_{LL,kV}$ = Line-to-line voltage rating, kV

A three-phase, 150-kVA transformer with a high-voltage winding of 12470GrdY/7200 V has a full-load current of 6.94 A on the primary (the same current as one 50-kVA single-phase transformer).

There are many types of three-phase connections used to serve three-phase load on distribution systems (ANSI/IEEE C57.105-1978; Long, 1984; Rusch

Five-legged wound core

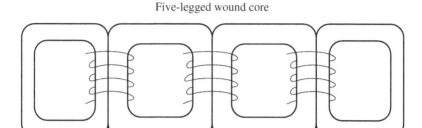

Four-legged stacked core

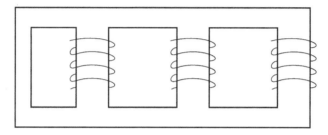

Three-legged stacked core

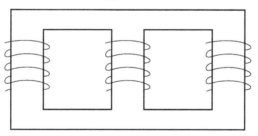

FIGURE 4.11
Three-phase core constructions.

and Good, 1989). Both the primary and secondary windings may be connected in different ways: delta, floating wye, or grounded wye. This notation describes the connection of the transformer windings, not the configuration of the supply system. A "wye" primary winding may be applied on a "delta" distribution system. On the primary side of three-phase distribution transformers, utilities have a choice between grounded and ungrounded winding connections. The tradeoffs are:

- *Ungrounded primary* — The delta and floating-wye primary connections are suitable for ungrounded and grounded distribution systems. Ferroresonance is more likely with ungrounded primary connections. Ungrounded primary connections do not supply ground fault current to the primary distribution system.

- *Grounded primary* — The grounded-wye primary connection is only suitable on four-wire grounded systems (either multigrounded or unigrounded). It is not for use on ungrounded systems. Grounded-wye primaries may provide an unwanted source for ground fault current.

Customer needs play a role in the selection of the secondary configuration. The delta configuration and the grounded-wye configuration are the two most common secondary configurations. Each has advantages and disadvantages:

- *Grounded-wye secondary* — Figure 4.12 shows the most commonly used transformers with a grounded-wye secondary winding: grounded wye – grounded wye and the delta – grounded wye. The

Grounded Wye -- Grounded Wye

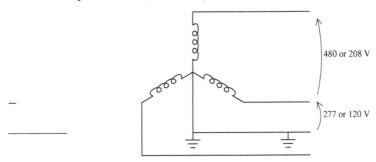

Delta -- Grounded Wye

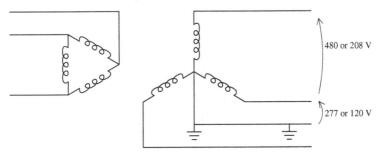

FIGURE 4.12
Three-phase distribution transformer connections with a grounded-wye secondary.

standard secondary voltages are 480Y/277 V and 208Y/120 V. The
480Y/277-V connection is suitable for driving larger motors; lighting
and other 120-V loads are normally supplied by dry-type transformers. A grounded-wye secondary adeptly handles single-phase loads
on any of the three phases with less concerns about unbalances.

- *Delta secondary* — An ungrounded secondary system like the delta
  can supply three-wire ungrounded service. Some industrial facilities
  prefer an ungrounded system, so they can continue to operate with
  line-to-ground faults. With one leg of the delta grounded at the
  midpoint of the winding, the utility can supply 240/120-V service.
  End-users can use more standard 230-V motors (without worrying
  about reduced performance when run at 208 V) and still run lighting
  and other single-phase loads. This tapped leg is often called the
  lighting leg (the other two legs are the *power* legs). Figure 4.13 shows
  the most commonly used connections with a delta secondary windings. This is commonly supplied with overhead transformers.

Many utilities offer a variety of three-phase service options and, of course,
most have a variety of existing transformer connections. Some utilities restrict
choices in an effort to increase consistency and reduce inventory. A restrictive
utility may only offer three choices: 480Y/277-V and 208Y/120-V four-wire,
three-phase services, and 120/240-V three-wire single-phase service.

For supplying customers requiring an ungrounded secondary voltage,
either a three-wire service or a four-wire service with 120 and 240 V, the
following provides the best connection:

- Floating wye – delta

For customers with a four-wire service, either of the following are normally
used:

- Grounded wye – grounded wye
- Delta – grounded wye

Choice of preferred connection is often based on past practices and equipment availability.

A wye – delta transformer connection shifts the phase-to-phase voltages
by 30° with the direction dependent on how the connection is wired. The
phase angle difference between the high-side and low-side voltage on delta
– wye and wye – delta transformers is 30°; by industry definition, the low
voltage lags the high voltage (IEEE C57.12.00-2000). Figure 4.14 shows
wiring diagrams to ensure proper phase connections of popular three-phase
connections.

Table 4.7 shows the standard winding designations shown on the nameplate of three-phase units.

Floating Wye -- Delta

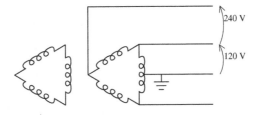

Common delta secondary connections:
240-V 3-wire
480-V 3-wire
240/120-V 4-wire (shown)

Delta -- Delta

Open Wye -- Open Delta

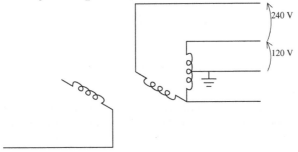

FIGURE 4.13
Common three-phase distribution transformer connections with a delta-connected secondary.

4.4.1   Grounded Wye – Grounded Wye

The most common three-phase transformer supply connection is the grounded wye – grounded wye connection. Its main characteristics are:

- *Supply* — Must be a grounded 4-wire system
- *Service*
  - Supplies grounded-wye service, normally either 480Y/277 V or 208Y/120 V.

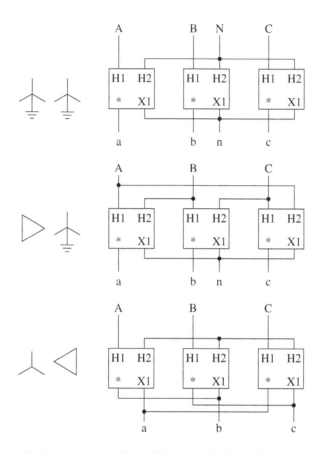

* is the opposite winding to X1, either X2, X3, or X4 depending on the transformer

FIGURE 4.14
Wiring diagrams for common transformer connections with additive units. Subtractive units have the same secondary connections, but the physical positions of X1 and * are reversed on the transformer.

- Cannot supply 120 and 240 V.
- Does not supply ungrounded service. (But a grounded wye – floating wye connection can.)
- *Tank heating* — Probable with three-legged core construction; less likely, but possible under severe unbalance with five-legged core construction. Impossible if made from three single-phase units.
- *Zero sequence* — All zero-sequence currents — harmonics, unbalance, and ground faults — transfer to the primary. It also acts as a high-impedance ground source to the primary.
- *Ferroresonance* — No chance of ferroresonance with a bank of single-phase units or triplex construction; some chance with a four- or five-legged core construction.

TABLE 4.7

Three-Phase Transformer Designations

Nomenclature	Examples	Description
E	2400	E shall indicate a winding that is permanently Δ connected for operation on an E volt system.
E_1Y	4160Y	E_1Y shall indicate a winding that is permanently Y connected without a neutral brought out (isolated) for operation on an E_1 volt system.
E_1Y/E	4160Y/2400	E_1Y/E shall indicate a winding that is permanently Y connected with a fully insulated neutral brought out for operation on an E_1 volt system, with E volts available from line to neutral.
E/E_1Y	2400/4160Y	E/E_1Y shall indicate a winding that may be Δ connected for operation on an E volt system, or may be Y connected without a neutral brought out (isolated) for operation on an E_1 volt system.
$E/E_1Y/E$	2400/4160Y/2400	$E/E_1Y/E$ shall indicate a winding that may be Δ connected for operation on an E volt system or may be Y connected with a fully insulated neutral brought out for operation on an E_1 volt system with E volts available from line to neutral.
E_1GrdY/E	12470GrdY/7200	E_1GrdY/E shall indicate a winding with reduced insulation and permanently Y connected, with a neutral brought out and effectively grounded for operation on an E_1 volt system with E volts available from line to neutral.
$E/E_1GrdY/E$	7200/12470GrdY/7200	$E/E_1GrdY/E$ shall indicate a winding, having reduced insulation, which may be Δ connected for operation on an E volt system or may be connected Y with a neutral brought out and effectively grounded for operation on an E_1 volt system with E volts available from line to neutral.
$V \times V_1$	7200 × 14400	$V \times V_1$ shall indicate a winding, the sections of which may be connected in parallel to obtain one of the voltage ratings (as defined in a–g) of V, or may be connected in series to obtain one of the voltage ratings (as defined in a–g) of V_1. Winding are permanently Δ or Y connected.

Source: IEEE Std. C57.12.00-2000. Copyright 2000 IEEE. All rights reserved.

- *Coordination* — Because ground faults pass through to the primary, larger transformer services and local protective devices should be coordinated with utility ground relays.

The grounded wye – grounded wye connection has become the most common three-phase transformer connection. Reduced ferroresonance is the main reason for the shift from the delta – grounded wye to the grounded wye – grounded wye.

Stray flux in the tank due to zero sequence current

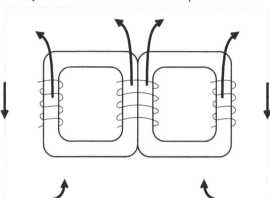

FIGURE 4.15
Zero-sequence flux caused by unbalanced voltages or unbalanced loads.

A grounded wye – grounded wye transformer with three-legged core construction is not suitable for supplying four-wire service. Unbalanced secondary loading and voltage unbalance on the primary system, these unbalances heat the transformer tank. In a three-legged core design, zero-sequence flux has no iron-core return path, so it must return via a high-reluctance path through the air gap and partially through the transformer tank (see Figure 4.15). The zero-sequence flux induces eddy currents in the tank that heat the tank.

A four- or five-legged core transformer greatly reduces the problem of tank heating with a grounded wye – grounded wye connection. The extra leg(s) provide an iron path for zero-sequence flux, so none travels into the tank. Although much less of a problem, tank heating can occur on four and five-legged core transformers under certain conditions; very large voltage unbalances may heat the tank. The outer leg cores normally do not have full capacity for zero-sequence flux (they are smaller than the inner leg cores), so under very high voltage unbalance, the outer legs may saturate. Once the legs saturate, some of the zero-sequence flux flows in the tank causing heating. The outer legs may saturate for a zero-sequence voltage of about 50 to 60% of the rated voltage. If a fuse or single-phase recloser or single-pole switch opens upstream of the transformer, the unbalance may be high enough to heat the tank, depending on the loading on the transformer and whether faults still exist. The worst conditions are when a single-phase interrupter clears a line-to-line or line-to-line-to-line fault (but not to ground) and the transformer is energized through one or two phases.

To completely eliminate the chance of tank heating, do not use a core-form transformer. Use a bank made of three single-phase transformers, or use triplex construction.

A wye – wye transformer with the primary and secondary neutrals tied together internally causes high line-to-ground secondary voltages if the neutral is not grounded. This connection cannot supply three-wire ungrounded service. Three-phase padmounted transformers with an H0X0 bushing have the neutrals bonded internally. If the H0X0 bushing is floated, high voltages can occur from phase to ground on the secondary.

To supply ungrounded secondary service with a grounded-wye primary, use a grounded wye – floating wye connection: the secondary should be floating wye with no connection between the primary and secondary neutral points.

4.4.2 Delta – Grounded Wye

The delta – grounded wye connection has several interesting features, many related to its delta winding, which establishes a local grounding reference and blocks zero-sequence currents from entering the primary.

- *Supply* — 3-wire or 4-wire system.
- *Service*
  - Supplies grounded-wye service, normally either 480Y/277 V or 208Y/120 V.
  - Cannot supply both 120 and 240 V.
  - Does not supply ungrounded service.
- *Ground faults* — This connection blocks zero sequence, so upstream ground relays are isolated from line-to-ground faults on the secondary of the customer transformer.
- *Harmonics* — The delta winding isolates the primary from zero-sequence harmonics created on the secondary. Third harmonics and other zero-sequence harmonics cannot get through to the primary (they circulate in the delta winding).
- *No primary ground source* — For line-to-ground faults on the primary, the delta – grounded wye connection cannot act as a grounding source.
- *Secondary ground source* — Provides a grounding source for the secondary, independent of the primary-side grounding configuration.
- *No tank heating* — The delta connection ensures that zero-sequence flux will not flow in the transformer's core. We can safely use a three-legged core transformer.
- *Ferroresonance* — Highly susceptible.

4.4.3 Floating Wye – Delta

The floating-wye – delta connection is popular for supplying ungrounded service and 120/240-V service. This type of connection may be used from either a grounded or ungrounded distribution primary. The main character-istics of this supply are:

- *Supply* — 3-wire or 4-wire system.
- *Service*
  - Can supply ungrounded service.
  - Can supply four-wire service with 240/120-V on one leg with a midtapped ground.
  - Cannot supply grounded-wye four-wire service.
- *Unit failure* — Can continue to operate if one unit fails if it is rewired as an open wye – open delta.
- *Voltage unbalance* — Secondary-side unbalances are more likely than with a wye secondary connection.
- *Ferroresonance* — Highly susceptible.

Do not use single-phase transformers with secondary breakers (CSPs) in this connection. If one secondary breaker opens, it breaks the delta on the secondary. Now, the primary neutral can shift wildly. The transformer may be severely overloaded by load unbalance or single phasing on the primary.

Facilities should ensure that single-phase loads only connect to the lighting leg; any miswired loads have overvoltages. The phase-to-neutral connection from the neutral to the opposite phase (where both power legs come together) is 208 V on a 240/120-V system.

The floating wye – delta is best used when supplying mainly three-phase load with a smaller amount of single-phase load. If the single-phase load is large, the three transformers making up the connection are not used as efficiently, and voltage unbalances can be high on the secondary.

In a conservative loading guideline, size the lighting transformer to supply all of the single-phase load plus 1/3 of the three-phase load (ANSI/IEEE C57.105-1978). Size each power leg to carry 1/3 of the three-phase load plus 1/3 of the single-phase load. ABB (1995) describes more accurate loading equations:

Lighting leg loading in kVA:

$$kVA_{bc} = \frac{1}{3}\sqrt{k_3^2 + 4k_1^2 + 4k_3 k_1 \cos\alpha}$$

Lagging power leg loading in kVA:

$$kVA_{ca} = \frac{1}{3}\sqrt{k_3^2 + k_1^2 - 2k_3 k_1 \cos(120° + \alpha)}$$

Leading power leg loading in kVA:

$$kVA_{ab} = \frac{1}{3}\sqrt{k_3^2 + k_1^2 - 2k_3 k_1 \cos(120° - \alpha)}$$

where

k_1 = single-phase load, kVA
k_3 = balanced three-phase load, kVA
$\alpha = \theta_3 - \theta_1$
θ_3 = phase angle in degrees for the three-phase load
θ_1 = phase angle in degrees for the single-phase load

For wye – delta connections, the wye on the primary is normally inten-tionally ungrounded. If it is grounded, it creates a grounding bank. This is normally undesirable because it may disrupt the feeder protection schemes and cause excessive circulating current in the delta winding. Utilities some-times use this connection as a grounding source or for other unusual reasons.

Delta secondary windings are more prone to voltage unbalance problems than a wye secondary winding (Smith et al., 1988). A balanced three-phase load can cause voltage unbalance if the impedances of each leg are different. With the normal practice of using a larger lighting leg, the lighting leg has a lower impedance. Voltage unbalance is worse with longer secondaries and higher impedance transformers. High levels of single-phase load also aggra-vate unbalances.

4.4.4 Other Common Connections

4.4.4.1 *Delta – Delta*

The main features and drawbacks of the delta – delta supply are:

- *Supply* — 3-wire or 4-wire system.
- *Service*
  - Can supply ungrounded service.
  - Can supply four-wire service with 240/120-V on one leg with a midtapped ground.
  - Cannot supply grounded-wye four-wire service.
- *Ferroresonance* — Highly susceptible.
- *Unit failure* — Can continue to operate if one unit fails (as an open delta – open delta).
- *Circulating current* — Has high circulating current if the turns ratios of each unit are not equal.

A delta – delta transformer may have high circulating current if any of the three legs has unbalance in the voltage ratio. A delta winding forms a series loop. Two windings are enough to fix the three phase-to-phase voltage vectors. If the third winding does not have the same voltage as that created by the series sum of the other two windings, large circulating currents flow to offset the voltage imbalance. ANSI/IEEE C57.105-1978 provides an example where the three phase-to-phase voltages summed to 1.5% of nominal as measured at the open corner of the delta winding (this voltage should be zero for no circulating current). With a 5% transformer impedance, a current equal to 10% of the transformer rating circulates in the delta when the open corner is connected. The voltage sees an impedance equal to the three winding impedances in series, resulting in a circulating current of 100% × 1.5% / (3×5%) = 10%. This circulating current directly adds to one of the three windings, possibly overloading the transformer.

Single-phase units with secondary breakers (CSPs) should not be used for the lighting leg. If the secondary breaker on the lighting leg opens, the load loses its neutral reference, but the phase-to-phase voltages are maintained by the other two legs (like an open delta – open delta connection). As with the loss of the neutral connection to a single-phase 120/240-V customer, unbalanced single-phase loads shift the neutral and create low voltages on one leg and high voltages on the lightly loaded leg.

4.4.4.2 Open Wye – Open Delta

The main advantage of the open wye – open delta transformer configuration is that it can supply three-phase load from a two-phase supply (but the supply must have a neutral).

The main features and drawbacks of the open wye – delta supply are:

- *Supply* — 2 phases and the neutral of a 4-wire grounded system.
- *Service*
  - Can supply ungrounded service.
  - Can supply four-wire service with 240/120-V on one leg with a midtapped ground.
  - Cannot supply grounded-wye four-wire service.
- *Ferroresonance* — Immune.
- *Voltage unbalance* — May have high secondary voltage unbalance.
- *Primary ground current* — Creates high primary-side current unbalance. Even with balanced loading, high currents are forced into the primary neutral.

Open wye – open delta connections are most efficiently applied when the load is predominantly single phase with some three-phase load, using one large lighting-leg transformer and another smaller unit. This connection is easily upgraded if the customer's three-phase load grows by adding a second power-leg transformer.

For sizing each bank, size the power leg for $1/\sqrt{3} = 0.577$ times the balanced three-phase load, and size the lighting leg for all of the single-phase load plus 0.577 times the three-phase load (ANSI/IEEE C57.105-1978). The following equations more accurately describe the split in loading on the two transformers (ABB, 1995). The load on the lighting leg in kVA is

$$kVA_L = \frac{k_3^2}{3} + k_1^2 + \frac{2k_3 k_1}{\sqrt{3}} \cos(\alpha + 30^\circ) \text{ for a leading lighting leg}$$

$$kVA_L = \frac{k_3^2}{3} + k_1^2 + \frac{2k_3 k_1}{\sqrt{3}} \cos(\alpha - 30^\circ) \text{ for a lagging lighting leg}$$

The power leg loading in kVA is:

$$kVA_L = \frac{k_3}{\sqrt{3}}$$

where
$\quad k_1$ = single-phase load, kVA
$\quad k_3$ = balanced three-phase load, kVA
$\quad \alpha = \theta_3 - \theta_1$
$\quad \theta_3$ = phase angle in degrees for the three-phase load
$\quad \theta_1$ = phase angle in degrees for the single-phase load

The lighting leg may be on the leading or lagging leg. In the open wye – open delta connection shown in Figure 4.13, the single-phase load is on the leading leg. For a lagging connection, switch the lighting and the power leg. Having the lighting connection on the leading leg reduces the loading on the lighting leg. Normally, the power factor of the three-phase load is less than that of the single-phase load, so α is positive, which reduces the loading on the lighting leg.

On the primary side, it is important that the two high-voltage primary connections are not made to the same primary phase. If this is accidentally done, the phase-to-phase voltage across the open secondary is two times the normal phase-to-phase voltage.

The open wye – open delta connection injects significant current into the neutral on the four-wire primary. Even with a balanced three-phase load, significant current is forced into the ground as shown in Figure 4.16. The extra unbalanced current can cause primary-side voltage unbalance and may trigger ground relays.

Open-delta secondary windings are very prone to voltage unbalance, which can cause excessive heating in end-use motors (Smith et al., 1988). Even balanced three-phase loads significantly unbalance the voltages. Voltage unbalance is less with lower-impedance transformers. Voltage unbalance

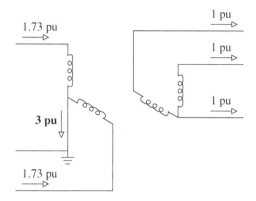

FIGURE 4.16
Current flow in an open wye – open delta transformer with balanced three-phase load.

reduces significantly if the connection is upgraded to a floating wye – closed delta connection. In addition, the component of the negative-sequence voltage on the primary (which is what really causes motor heating) can add to that caused by the transformer configuration to sometimes cause a negative-sequence voltage above 5% (which is a level that significantly increases heating in a three-phase induction motor).

While an unusual connection, it is possible to supply a balanced, grounded four-wire service from an open-wye primary. This connection (open wye – partial zig-zag) can be used to supply 208Y/120-V service from a two-phase line. One of the 120/240-V transformers must have four bushings; X2 and X3 are not tied together but connected as shown in Figure 4.17. Each of the transformers must be sized to supply 2/3 of the balanced three-phase load. If four-bushing transformers are not available, this connection can be made with three single-phase transformers. Instead of the four-bushing transformer, two single-phase transformers are placed in parallel on the primary, and the secondary terminals of each are configured to give the arrangement in Figure 4.17.

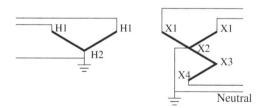

FIGURE 4.17
Quasiphasor diagram of an open-wye primary connection supplying a wye four-wire neutral service such as 208Y/120 V. (From ANSI/IEEE Std. C57.105-1978. Copyright 1978 IEEE. All rights reserved.)

4.4.4.3 Other Suitable Connections

While not as common, several other three-phase connections are used at times:

- *Open delta – Open delta* — Can supply a three-wire ungrounded service or a four-wire 120/240-V service with a midtapped ground on one leg of the transformer. The ungrounded high-side connection is susceptible to ferroresonance. Only two transformers are needed, but it requires all three primary phases. This connection is less efficient for supplying balanced three-phase loads; the two units must total 115% of the connected load. This connection is most efficiently applied when the load is predominantly single phase with some three-phase load, using one large lighting-leg transformer and another smaller unit.

- *Delta – Floating wye* — Suitable for supplying a three-wire ungrounded service. The ungrounded high-side connection is susceptible to ferroresonance.

- *Grounded wye – Floating wye* — Suitable for supplying a three-wire ungrounded service from a multigrounded primary system. The grounded primary-side connection reduces the possibility of ferroresonance.

4.4.5 Neutral Stability with a Floating Wye

Some connections with a floating-wye winding have an unstable neutral, which we should avoid. Unbalanced single-phase loads on the secondary, unequal magnetizing currents, and suppression of third harmonics — all can shift the neutral.

Consider a *floating wye – grounded wye* connection. In a wye – wye transformer, the primary and secondary voltages have the same vector relationships. The problem is that the neutral point does not have a grounding source; it is free to float. Unbalanced loads or magnetizing currents can shift the neutral and create high neutral-to-earth voltages and overvoltages on the phases with less loading. The reverse connection with a grounded wye – floating wye works because the primary-side neutral is connected to the system neutral, which has a grounding source. The grounding source fixes the neutral voltage.

In a floating wye, current in one branch is dependent on the currents in the other two branches. What flows in one branch must flow out the other two branches. This creates conditions that shift the neutral (Blume et al., 1951):

- *Unbalanced loads* — Unequal single-phase loads shift the neutral point. Zero-sequence current has no path to flow (again, the ground source is missing). Loading one phase drops the voltage on that phase and raises the voltage on the other two phases. Even a small unbalance significantly shifts the neutral.

- *Unequal magnetizing currents* — Just like unequal loads, differences in the amount of magnetizing current each leg needs can shift the floating neutral. In a four- or five-legged core, the asymmetry of the core causes unequal magnetizing requirements on each phase.

- *Suppression of third harmonics* — Magnetizing currents contain significant third harmonics that are zero sequence. But, the floating wye connection has no ground source to absorb the zero-sequence currents, so they are suppressed. The suppression of the zero-sequence currents generates a significant third-harmonic voltage in each winding, about 50% of the phase voltage on each leg according to Blume et al. (1951). With the neutral grounded in the floating wye – grounded wye, a significant third-harmonic voltage adds to each phase-to-ground load. If the neutral is floating (on the wye–wye transformer with the neutrals tied together), the third-harmonic voltage appears between the neutral and ground.

In addition to the floating wye – grounded wye, avoid these problem connections that have an unstable neutral:

- *Grounded wye – grounded wye on a three-wire system* — The grounded-wye on the primary does not have an effective grounding source, so it acts the same as a floating-wye – grounded-wye.

- *A wye – wye transformer with the primary and secondary neutrals tied together internally (the H0X0 bushing) but with the neutral left floating* — Again, the neutral point can float. Unbalanced loading is not a problem, but magnetizing currents and suppression of third harmonics are. These can generate large voltages between the neutral point and ground (and between the phase wires and ground). If the secondary neutral is isolated from the primary neutral, each neutral settles to a different value. But when the secondary neutral is locked into the primary neutral, the secondary neutral follows the neutral shift of the primary and shifts the secondary phases relative to ground.

Another poor connection is the floating wye – floating wye. Although not as bad as the floating-wye – grounded-wye connection, the neutral can shift if the connection is made of three units of different magnetizing characteristics. The neutral shift can lead to an overvoltage across one of the windings. Also, high harmonic voltage appears on the primary-side neutral (which is okay if the neutral is properly insulated from the tank).

Three-legged core transformers avoid some of the problems with a floating wye. The phantom tertiary acts as a mini ground source, stabilizes the neutral, and even allows some unbalance of single-phase loads. But as it stabilizes the neutral, the unbalances heat the tank. Given that, it is best to avoid these transformer connections. They provide no features or advantages over other transformer connections.

4.4.6 Sequence Connections of Three-Phase Transformers

The connection determines the effect on zero sequence, which impacts unbalances and response to line-to-ground faults and zero-sequence harmonics. Figure 4.18 shows how to derive sequence connections along with common examples. In general, three-phase transformers may affect the zero-sequence circuit as follows:

- *Isolation* — A floating wye – delta connection isolates the primary from the secondary in the zero sequence.

- *Pass through* — The grounding of the grounded wye – grounded wye connection is determined by the grounding upstream of the transformer.

- *Ground source* — A delta – grounded wye connection provides a ground source on the secondary. (And, the delta – grounded wye connection also isolates the primary from the secondary.)

4.5 Loadings

Distribution transformers are *output rated*; they can deliver their rated kVA without exceeding temperature rise limits when the following conditions apply:

- The secondary voltage and the voltage/frequency do not exceed 105% of rating. So, a transformer is a constant kVA device for a voltage from 100 to 105% (the standards are unclear below that, so treat them as constant current devices).
- The load power factor $\geq$ 80%.
- Frequency $\geq$ 95% of rating.

The transformer loading and sizing guidelines of many utilities are based on ANSI/IEEE C57.91-1981.

Modern distribution transformers are 65°C rise units, meaning they have normal life expectancy when operated with an average winding temperature rise above ambient of not more than 65°C and a hottest spot winding temperature rise of not more than 80°C. Some older units are 55°C rise units, which have less overloading capability.

At an ambient temperature of 30°C, the 80°C hottest-spot rise for 65°C rise units gives a hottest-spot winding temperature of 110°C. The hot-spot tem-

Zero-sequence diagram

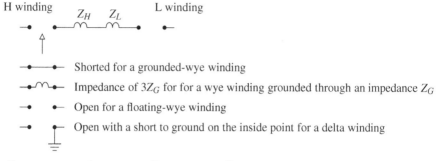

Shorted for a grounded-wye winding

Impedance of $3Z_G$ for for a wye winding grounded through an impedance Z_G

Open for a floating-wye winding

Open with a short to ground on the inside point for a delta winding

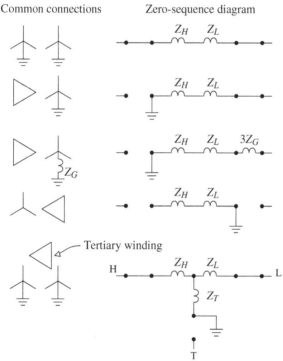

Common connections   Zero-sequence diagram

3-legged core (acts as a high-impedance tertiary)

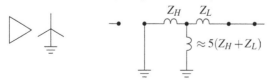

FIGURE 4.18
Zero-sequence connections of various three-phase transformer connections.

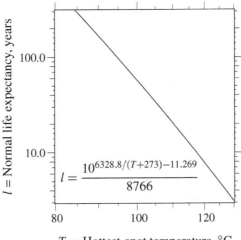

$$l = \frac{10^{6328.8/(T+273)-11.269}}{8766}$$

T = Hottest-spot temperature, °C

FIGURE 4.19

Transformer life as a function of the hottest-spot winding temperature.

perature on the winding is critical; that's where insulation degrades. The insulation's life exponentially relates to hot-spot winding temperature as shown in Figure 4.19. At 110°C, the normal life expectancy is 20 years. Because of daily and seasonal load cycles, most of the time temperatures are nowhere near these values. Most of the time, temperatures are low enough not to do any significant insulation degradation. We can even run at temperatures above 110°C for short periods. For the most economic operation of distribution transformers, they are normally sized to operate at significant overloads for short periods of the year.

We can load distribution transformers much more heavily when it is cold. Locations with winter-peaking loads can have smaller transformers for a given loading level. The transformer's kVA rating is based on an ambient temperature of 30°C. For other temperatures, ANSI/IEEE C57.91-1981 suggests the following adjustments to loading capability:

- > 30°C: decrease loading capability by 1.5% of rated kVA for each °C above 30°C.

- < 30°C: increase loading capability by 1% of rated kVA for each °C above 30°C.

Ambient temperature estimates for a given region can be found using historical weather data. For loads with normal life expectancy, ANSI/IEEE C57.91-1981 recommends the following estimate of ambient temperature:

- *Average daily temperature for the month involved* — As an approximation, the average can be approximated as the average of the daily highs and the daily lows.

For short-time loads where we are designing for a moderate sacrifice of life, use:

- *Maximum daily temperature*

In either case, the values should be averaged over several years for the month involved. C57.91-1981 also suggests adding 5°C to be conservative. These values are for outdoor overhead or padmounted units. Transformers installed in vaults or other cases with limited air flow may require some adjustments.

Transformers should also be derated for altitudes above 3300 ft (1000 m). At higher altitudes, the decreased air density reduces the heat conducted away from the transformer. ANSI/IEEE C57.91-1981 recommends derating by 0.4% for each 330 ft (100 m) that the altitude is above 3300 ft (1000 m).

Load cycles play an important role in determining loading. ANSI/IEEE C57.91-1981 derives an equivalent load cycle with two levels: the peak load and the initial load. The equivalent two-step load cycle may be derived from a more detailed load cycle. The guide finds a continuous load and a short-duration peak load. Both are found using the equivalent load value from a more complicated load cycle:

$$L = \sqrt{\frac{L_1^2 t_1 + L_2^2 t_2 + L_3^2 t_3 + \cdots + L_n^2 t_n}{t_1 + t_2 + t_3 + \cdots + t_n}}$$

where
L = equivalent load in percent, per unit, or actual kVA
$L_1, L_2, \ldots,$ = The load steps in percent, per unit, or actual kVA
$t_1, t_2, \ldots,$ = The corresponding load durations

The continuous load is the equivalent load found using the equation above for 12 h preceding and 12 h following the peak and choosing the higher of these two values. The guide suggests using 1-h time blocks. The peak is the equivalent load from the equation above where the irregular peak exists.

The C57.91 guide has loading guidelines based on the peak duration and continuous load prior to the peak. Table 4.8 shows that significant overloads are allowed depending on the preload and the duration of the peak.

Because a region's temperature and loading patterns vary significantly, there is no universal transformer application guideline. Coming up with standardized tables for initial loading is based on a prediction of peak load, which for residential service normally factors in the number of houses, average size (square footage), central air conditioner size, and whether electric heat is used. Once the peak load is estimated, it is common to pick a transformer with a kVA rating equal to or greater than the peak load kVA estimate. With this arrangement, some transformers may operate significantly above their ratings for short periods of the year. Load growth can push the peak

TABLE 4.8

Transformer Loading Guidelines

Peak Load Duration, Hours	Extra Loss of Life[a], %	Equivalent Peak Loading in Per Unit of Rated kVA with the Percent Preload and Ambient Temperatures Given Below										
		50% Preload Ambient Temp., °C				75% Preload Ambient Temp., °C				90% Preload Ambient Temp., °C		
		20	30	40	50	20	30	40	50	20	30	40
1	Normal	2.26	2.12	1.96	1.79	2.12	1.96	1.77	1.49	2.02	1.82	1.43
	0.05	2.51	2.38	2.25	2.11	2.40	2.27	2.12	1.95	2.31	2.16	1.97
	0.10	2.61	2.49	2.36	2.23	2.50	2.37	2.22	2.07	2.41	2.27	2.11
	0.50	2.88	2.76	2.64	2.51	2.77	2.65	2.52	2.39	2.70	2.57	2.43
2	Normal	1.91	1.79	1.65	1.50	1.82	1.68	1.52	1.26	1.74	1.57	1.26
	0.05	2.13	2.02	1.89	1.77	2.05	1.93	1.80	1.65	1.98	1.85	1.70
	0.10	2.22	2.10	1.99	1.87	2.14	2.02	1.90	1.75	2.07	1.95	1.81
	0.50	2.44	2.34	2.23	—	2.37	2.26	2.15	—	2.31	2.20	2.08
4	Normal	1.61	2.50	1.38	1.25	1.56	1.44	1.30	1.09	1.50	1.36	1.13
	0.05	1.80	1.70	1.60	1.48	1.76	1.65	1.54	1.40	1.71	1.60	1.47
	0.10	1.87	1.77	1.67	—	1.83	1.72	1.62	1.50	1.79	1.68	1.56
	0.50	2.06	1.97	—	—	2.02	1.93	—	—	1.99	1.89	—
8	Normal	1.39	1.28	1.18	1.05	1.36	1.25	1.13	0.96	1.33	1.21	1.02
	0.05	1.55	1.46	1.36	1.25	1.53	1.43	1.33	1.21	1.51	1.41	1.29
	0.10	1.61	1.53	1.43	1.33	1.59	1.50	1.41	1.30	1.57	1.47	1.38
	0.50	1.78	1.69	1.61	—	1.76	1.67	1.58	—	1.74	1.65	1.56
24	Normal	1.18	1.08	0.97	0.86	1.17	1.07	0.97	0.84	1.16	1.07	0.95
	0.05	1.33	1.24	1.15	1.04	1.33	1.24	1.13	1.04	1.32	1.23	1.13
	0.10	1.39	1.30	1.21	1.11	1.38	1.29	1.20	1.10	1.38	1.29	1.20
	0.50	1.54	1.45	1.37	1.28	1.53	1.45	1.37	1.28	1.53	1.45	1.36

[a] Extra loss of life in addition to 0.0137% per day loss of life for normal life expectancy.

Source: ANSI/IEEE C57.91-1981, *IEEE Guide for Loading Mineral-Oil-Immersed Overhead and Pad-Mounted Distribution Transformers Rated 500 kVA and Less with 65 Degrees C or 55 Degrees C Average Winding Rise.*

load above the peak kVA estimate, and inaccuracy of the load prediction will mean that some units are going to be loaded more than expected. The load factor (the ratio of average demand to peak demand) for most distribution transformers is 40 to 60%. Most distribution transformers are relatively lightly loaded most of the time, but some have peak loads well above their rating. In analysis of data from three utilities, the Oak Ridge National Laboratory found that distribution transformers have an average load of 15 to 40% of their rating, with 30% being most typical (ORNL-6925, 1997).

The heat input into the transformer is from no-load losses and from load losses. The economics of transformer application and purchasing involve consideration of the thermal limitations as well as the operating costs of the losses. Transformer stocking considerations also play a role. For residential customers, a utility may limit inventory to 15, 25, and 50-kVA units (5, 10, 15, 25, 37.5, 50-kVA units are standard sizes).

Some utilities use transformer load management programs to more precisely load transformers to get the most economic use of each transformer's

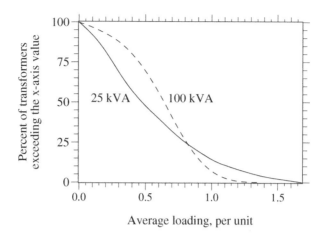

FIGURE 4.20
Distributions of average loadings of two transformer sizes at one utility. (From [ORNL-6927, 1998])

life. These programs take billing data for the loads from each transformer to estimate that transformer's loading. These programs allow the utility to more aggressively load transformers because those needing changeout can be targeted more precisely. Load management programs require data setup and maintenance. Most important, each meter must be tied to a given transformer (many utilities have this information infrastructure, but some do not).

Transformer loadings vary considerably. Figure 4.20 shows the distribution of average loadings on two sizes of transformers at one typical utility. Most transformers are not heavily loaded: in this case, 85% of units have average loadings less than the nameplate. Many units are very lightly loaded, and 10% are quite heavily loaded. Smaller units have more spread in their loading.

Seevers (1995) demonstrates a simple approach to determining transformer loading. Their customers (in the southern U.S.) had 1 kW of demand for every 400 kWh's, regardless of whether the loads peaked in the winter or summer. Seevers derived the ratio by comparing substation demand with kWh totals for all customers fed from the substation (after removing primary-metered customers and other large loads). To estimate the load on a given transformer, sum the kWh for the month of highest usage for all customers connected to the transformer and convert to peak demand, in this case by dividing by 400 kWh per kW-demand. While simple, this method identifies grossly undersized or oversized transformers. Table 4.9 shows guidelines for replacement of underloaded transformers.

Transformers with an internal secondary breaker (CSPs) are a poor-man's form of transformer load management. If the breaker trips from overload, replace the transformer (unless there are extraordinary weather and loading conditions that are unlikely to be repeated).

TABLE 4.9

One Approach to a Transformer Replacement Program

Existing Transformer kVA	Loading Estimate in kVA	Recommended Size in kVA
25	10 or less	10
37.5	15 or less	10
50	20 or less	15
75	37.5 or less	37.5
100	50 or less	50
167	100 or less	100
	75 or less	75 or 50

Source: Seevers, O. C., *Management of Transmission and Distribution Systems*, Penn Well Publishing Company, Tulsa, OK, 1995.

Especially in high-lightning areas, consider the implications of reduction of insulation capability. At hottest-spot temperatures above 140°C, the solid insulation and the oil may release gasses. While not permanently reducing insulation, the short-term loss of insulation strength can make the transformer susceptible to damage from lightning-caused voltage surges. The thermal time constant of the winding is very short, 5 to 15 min. On this time scale, loads on distribution transformers are quite erratic with large, short-duration overloads (well above the 20- or 30-min demand loadings). These loads can push the winding hottest-spot temperature above 140°C.

Padmounted transformers have a special concern related to loading: case temperatures. Under heavy loading on a hot day, case temperatures can become hot. ABB measured absolute case temperatures of 185 to 200°F (85 to 95°C) and case temperature rises above ambient of 50 to 60°C on 25 and 37.5-kVA transformers at 180% loadings and on a 50-kVA transformer at 150% continuous load (NRECA RER Project 90-8, 1993). The hottest temperatures were on the sides of the case where the oil was in contact with the case (the top of the case was significantly cooler). While these temperatures sound quite high, a person's pain-withdrawal reflexes will normally protect against burns for normal loadings that would be encountered. Reflexes will protect against blistering and burning for case temperatures below 300°F (149°C). Skin contacts must be quite long before blistering occurs. For a case temperature of 239°F (115°C), NRECA reported that the skin-contact time to blister is 6.5 sec (which is more than enough time to pull away). At 190°F (88°C), the contact time to blister is 19 sec.

4.6 Losses

Transformer losses are an important purchase criteria and make up an appreciable portion of a utility's overall losses. The Oak Ridge National Laboratory estimates that distribution transformers account for 26% of transmission and distribution losses and 41% of distribution and subtransmission losses (ORNL-6804/R1, 1995). At one utility, Grainger and Kendrew (1989) estimated that distribution transformers were 55% of distribution losses and 2.14% of electricity sales; of the two main contributors to losses, 86% were no-load losses, and 14% were load losses.

Load losses are also called copper or wire or winding losses. Load losses are from current through the transformer's windings generating heat through the winding resistance as I^2R.

No-load losses are the continuous losses of a transformer, regardless of load. No-load losses for modern silicon-steel-core transformers average about 0.2% of the transformer rating (a typical 50-kVA transformer has no-load losses of 100 W), but designs vary from 0.15 to 0.4% depending on the needs of the utility. No-load losses are also called iron or core losses because they are mainly a function of the core materials. The two main components of no-load losses are eddy currents and hysteresis. Hysteresis describes the memory of a magnetic material. More force is necessary to demagnetize magnetic material than it takes to magnetize it; the magnetic domains in the material resist realignment. Eddy current losses are small circulating currents in the core material. The steel core is a conductor that carries an alternating magnetic field, which induces circulating currents in the core. These currents through the resistive conductor generate heat and losses. Cores are typically made from cold-rolled, grain-oriented silicon steel laminations. Manufacturers limit eddy currents by laminating the steel core in 9- to 14-mil thick layers, each insulated from the other. Core losses increase with steady-state voltage.

Hysteresis losses are a function of the volume of the core, the frequency, and the maximum flux density (Sankaran, 2000):

$$P_h \propto V_e f B^{1.6}$$

where
  V_e = volume of the core
  f = frequency
  B = maximum flux density

The eddy-current losses are a function of core volume, frequency, flux density, lamination thicknesses, and resistivity of the core material (Sankaran, 2000):

$$P_e \propto V_e B^2 f^2 t^2 / r$$

where
  t = thickness of the laminations
  r = resistivity of the core material

TABLE 4.10

Loss Reduction Alternatives

	No-Load Losses	Load Losses	Cost
To Decrease No-Load Losses			
Use lower-loss core materials	Lower	No change[a]	Higher
Decrease flux density by:			
(1) increasing core CSA[b]	Lower	Higher	Higher
(2) decreasing volts/turn	Lower	Higher	Higher
Decrease flux path length by decreasing conductor CSA	Lower	Higher	Lower
To Decrease Load Losses			
Use lower-loss conductor materials	No change	Lower	Higher
Decrease current density by increasing conductor CSA	Higher	Lower	Higher
Decrease current path length by:			
(1) decreasing core CSA	Higher	Lower	Lower
(2) increasing volts/turn	Higher	Lower	Higher

[a] Amorphous core materials would result in higher load losses.
[b] CSA=cross-sectional area

Source: ORNL-6847, *Determination Analysis of Energy Conservation Standards for Distribution Transformers*, Oak Ridge National Laboratory, U.S. Department of Energy, 1996.

Amorphous core metals significantly reduce core losses — as low as one quarter of the losses of silicon-steel cores — on the order of 0.005 to 0.01% of the transformer rating. Amorphous cores do not have a crystalline structure like silicon-steel cores; their iron atoms are randomly distributed. Amorphous materials are made by rapidly cooling a molten alloy, so crystals do not have a chance to form. Such core materials have low hysteresis loss. Eddy current-losses are very low because of the high resistivity of the material and very thin laminations (1-mil thick). Amorphous-core transformers are larger for the same kVA rating and have higher initial costs.

Load losses, no-load losses, and purchase price are all interrelated. Approaches to reduce load losses tend to increase no-load losses and vice versa. For example, a larger core cross-sectional area decreases no-load losses (the flux density core is less), but this requires longer winding conductors and more I^2R load losses. Table 4.10 shows some of the main tradeoffs.

Information from transformer load management programs can help with transformer loss analysis. Table 4.11 shows typical transformer loading data from one utility. The average load on most transformers is relatively low (25 to 30% of transformer rating), which highlights the importance of no-load losses. The total equivalent losses on a transformer are

$$L_{total} = P^2 F_{ls} L_{load} + L_{no-load}$$

where

L_{total} = average losses, kW (multiply this by 8760 to find the annual kilo-watt-hours)

P = peak transformer load, per unit

F_{ls} = loss factor, per unit

$L_{no-load}$ = rated no-load losses, kW

L_{load} = rated load losses, kW

Many utilities evaluate the total life-cycle cost of distribution transformers, accounting for the initial purchase price and the cost of losses over the life of the transformer (the total owning cost or TOC). The classic work done by Gangel and Propst (1965) on transformer loads and loss evaluation provides the foundation for much of the later work. Many utilities follow the Edison Electric Association's economic evaluation guidelines (EEI, 1981). To evaluate the total owning cost, the utility's cost of losses are evaluated using transformer loading assumptions, including load factor, coincident factor, and responsibility factor. Utilities typically assign an equivalent present value for the costs of no-load losses and another for the cost of load losses. Loss values typically range from $2 to $4/W of no-load losses and $0.50 to $1.50/W of load losses (ORNL-6847, 1996). Utilities that evaluate the life costs of transformers purchase lower-loss transformers. For example, a 50-

TABLE 4.11

Summary of the Loading of One Utility's Single-Phase Pole-Mounted Distribution Transformers

Size (kVA)	No. of Installed Transformers	MWh/ Transformer	Annual PU Avg. Load	Annual PU Load Factor	Calculated Loss Factor
10	59,793	21	0.267	0.405	0.200
15	106,476	34	0.292	0.430	0.221
25	118,584	60	0.309	0.444	0.234
37	77,076	96	0.329	0.445	0.235
50	50,580	121	0.308	0.430	0.222
75	24,682	166	0.281	0.434	0.225
100	8,457	220	0.280	0.463	0.252
167	3,820	372	0.283	0.516	0.304
250	592	631	0.320	0.568	0.360
333	284	869	0.331	0.609	0.407
500	231	1,200	0.304	0.598	0.394
667	9	1,666	0.317	0.476	0.264
833	51	2,187	0.333	0.629	0.431

Note: PU = per unit

Source: ORNL-6925, *Supplement to the "Determination Analysis" (ORNL-6847) and Analysis of the NEMA Efficiency Standard for Distribution Transformers,* Oak Ridge National Laboratory, U.S. Department of Energy, 1997.

kVA single-phase, non-loss-evaluated transformer would have approximately 150 W of no-load losses and 675 W of load losses; the same loss-evaluated transformer would have approximately 100 W of no-load losses and 540 W of load losses (ORNL-6925, 1997). Nickel (1981) describes an economic approach in detail and compares it to the EEI method. The IEEE has developed a more recent guide (C57.12.33).

4.7   Network Transformers

Network transformers, the distribution transformers that serve grid and spot networks, are large three-phase units. Network units are normally vault-types or subway types, which are defined as (ANSI C57.12.40-1982):

- *Vault-type transformers* — Suitable for occasional submerged operation
- *Subway-type transformers* — Suitable for frequent or continuous submerged operation

Network transformers are often housed in vaults. Vaults are underground rooms accessed through manholes that house transformers and other equipment. Vaults may have sump pumps to remove water, air venting systems, and even forced-air circulation systems. Network transformers are also used in buildings, usually in the basement. In these, vault-type transformers may be used (as long as the room is properly built and secured for such use). Utilities may also use dry-type units and units with less flammable insulating oils.

A network transformer has a three-phase, primary-side switch that can open, close, or short the primary-side connection to ground. The standard secondary voltages are 216Y/125 V and 480Y/277 V. Table 4.12 shows standard sizes. Transformers up to 1000 kVA have a 5% impedance; above 1000 kVA, 7% is standard. X/R ratios are generally between 3 and 12. Lower impedance transformers (say 4%) have lower voltage drop and higher secondary fault currents. (Higher secondary fault currents help on a network to burn clear faults.) Lower impedance has a price though — higher circulating currents and less load balance between transformers. Network trans-

TABLE 4.12

Standard Network Transformer Sizes

Standard Ratings, kVA	
216Y/125 V	300, 500, 750, 1000
480Y/277 V	500, 750, 1000, 1500, 2000, 2500

formers may also be made out of standard single-phase distribution transformers, but caution is warranted if the units have very low leakage impedances (which could cause very high circulating currents and secondary fault levels higher than network protector ratings).

Most network transformers are connected delta – grounded wye. By blocking zero sequence, this connection keeps ground currents low on the primary cables. Then, we can use a very sensitive ground-fault relay on the substation breaker. Blocking zero sequence also reduces the current on cable neutrals and cable sheaths, including zero-sequence harmonics, mainly the third harmonic. One disadvantage of this connection is with combination feeders — those that feed network loads as well as radial loads. For a primary line-to-ground fault, the feeder breaker opens, but the network transformers will continue to backfeed the fault until all of the network protectors operate (and some may stick). Now, the network transformers backfeed the primary feeder as an ungrounded circuit. An ungrounded circuit with a single line-to-ground fault on one phase causes a neutral shift that raises the line-to-neutral voltage on the unfaulted phases to line-to-line voltage. The non-network load connected phase-to-neutral is subjected to this overvoltage.

Some networks use grounded wye – grounded wye connections. This connection fits better for combination feeders. For a primary line-to-ground fault, the feeder breaker opens. Backfeeds to the primary through the network still have a grounding reference with the wye – wye connection, so chances of overvoltages are limited. The grounded wye – grounded wye connection also reduces the change of ferroresonance in cases where a transformer has single-pole switching.

Most network transformers are core type, either a three- or five-legged core. The three-legged core, either with a stacked or wound core, is suitable for a delta – grounded wye connection (but not a grounded wye – grounded wye connection because of tank heating). A five-legged core transformer is suitable for either connection type.

4.8 Substation Transformers

In a distribution substation, power-class transformers provide the conversion from subtransmission circuits to the distribution primary. Most are connected delta – grounded wye to provide a ground source for the distribution neutral and to isolate the distribution ground system from the subtransmission system.

Station transformers can range from 5 MVA in smaller rural substations to over 80 MVA at urban stations (base ratings). Stations with two banks, each about 20 MVA, are common. Such a station can serve about six to eight feeders.

Power transformers have multiple ratings, depending on cooling methods. The base rating is the self-cooled rating, just due to the natural flow to the surrounding air through radiators. The transformer can supply more load with extra cooling turned on. Normally, fans blow air across the radiators and/or oil circulating pumps. Station transformers are commonly supplied with OA/FA/FOA ratings. The OA is open air, FA is forced air cooling, and FOA is forced air cooling plus oil circulating pumps.

The ANSI ratings were revised in the year 2000 to make them more consistent with IEC designations. This system has a four-letter code that indicates the cooling (IEEE C57.12.00-2000):

- *First letter* — Internal cooling medium in contact with the windings:
  - **O** mineral oil or synthetic insulating liquid with fire point = 300°C
  - **K** insulating liquid with fire point > 300°C
  - **L** insulating liquid with no measurable fire point
- *Second letter* — Circulation mechanism for internal cooling medium:
  - **N** natural convection flow through cooling equipment and in windings
  - **F** forced circulation through cooling equipment (i.e., coolant pumps); natural convection flow in windings (also called nondirected flow)
  - **D** forced circulation through cooling equipment, directed from the cooling equipment into at least the main windings
- *Third letter* — External cooling medium:
  - **A** air
  - **W** water
- *Fourth letter* — Circulation mechanism for external cooling medium:
  - **N** natural convection
  - **F** forced circulation: fans (air cooling), pumps (water cooling)

So, OA/FA/FOA is equivalent to ONAN/ONAF/OFAF. Each cooling level typically provides an extra one-third capability: 21/28/35 MVA. Table 4.13 shows equivalent cooling classes in the old and new naming schemes.

Utilities do not overload substation transformers as much as distribution transformers, but they do run them hot at times. As with distribution transformers, the tradeoff is loss of life versus the immediate replacement cost of the transformer. Ambient conditions also affect loading. Summer peaks are much worse than winter peaks. IEEE Std. C57.91-1995 provides detailed loading guidelines and also suggests an approximate adjustment of 1% of the maximum nameplate rating for every degree C above or below 30°C. The hottest spot conductor temperature is the critical point where insulation degrades. Above a hot-spot conductor temperature of 110°C, life expectancy decreases exponentially. The life halves for every 8°C increase in operating temperature. Most of the time, the hottest temperatures are nowhere near

TABLE 4.13

Equivalent Cooling Classes

Year 2000 Designations	Designations Prior to Year 2000
ONAN	OA
ONAF	FA
ONAN/ONAF/ONAF	OA/FA/FA
ONAN/ONAF/OFAF	OA/FA/FOA
OFAF	FOA
OFWF	FOW

Source: IEEE Std. C57.12.00-2000. Copyright 2000 IEEE. All rights reserved.

this. Tillman (2001) provides the loading guide for station transformers shown in Table 4.14.

The impedance of station transformers is normally about 7 to 10%. This is the impedance on the base rating, the self-cooled rating (OA or ONAN). The impedance is normally higher for voltages on the high-side of the transformer that are higher (like 230 kV). Transformer impedance can be specified when ordering. Large stations with 50 plus MVA transformers are normally provided with extra impedance to control fault currents, some as high as 30% on the transformer's base rating.

The positive and zero-sequence impedances are the same for a shell-type transformer, so the bolted fault currents on the secondary of the transformer are the same for a three-phase fault and for a line-to-ground fault (provided that both are fed from an infinite bus). In a three-legged core type transformer, the zero-sequence impedance is lower than the positive-sequence impedance (typically $Z_0 = 0.85Z_1$), so ground faults can cause higher currents. With a three-legged core transformer design, there is no path for zero-sequence flux. Therefore, zero-sequence current will meet a lower-impedance branch. This makes the core-type transformer act as if it had a delta-connected tertiary winding. This is the magnetizing branch (from line to ground), and this effectively reduces the zero-sequence impedance. In a shell-type transformer, there is a path through the iron for flux to flow, so the excitation impedance to zero sequence is high.

Because most distribution circuits are radial, the substation transformer is a critical component. Power transformers normally have a failure rate between 1 to 2% annually (CEA 485 T 1049, 1996; CIGRE working group 12.05, 1983; IEEE Std. 493-1997). Many distribution stations are originally designed with two transformers, where each is able to serve all of the substation's feeders if one of the transformers fails. Load growth in some areas has severely reduced the ability of one transformer to supply the whole station. To ensure transformer reliability, use good lightning protection and thermal management. Do not use reduced-BIL designs (BIL is the basic lightning impulse insulation level). Also, reclosing and relaying practices should ensure that excessive through faults do not damage transformers.

TABLE 4.14

Example Substation Transformer Loading Guide

Type of Load	FA (ONAF) Max Top Oil Temp (°C)	NDFOA (OFAF) Max Top Oil Temp (°C)	Max Winding Temp (°C)	Max % Load
Normal summer load	105	95	135	130
Normal winter load	80	70	115	140
Emergency summer load	115	105	150	140
Emergency winter load	90	80	130	150
Non-cyclical load	95	85	115	110

Alarm Settings	FA 65°C Rise	NDFOA 65°C Rise
Top Oil	105°C	95°C
Hot Spot	135°C	135°C
Load Amps	130%	130%

Notes: (1) The normal summer loading accounts for periods when temperatures are abnormally high. These might occur every 3 to 5 years. For every degree C that the normal ambient temperature during the hottest month of the year exceeds 30°C, de-rate the transformer 1% (i.e., 129% loading for 31°C average ambient). (2) The % load is given on the basis of the current rating. For MVA loading, multiply by the per unit output voltage. If the output voltage is 0.92 per unit, the recommended normal summer MVA loading is 120%. (3) Exercise caution if the load power factor is less than 0.95 lagging. If the power factor is less than 0.92 lagging, then lower the recommended loading by 10% (i.e., 130 to 120%). (4) Verify that cooling fans and pumps are in good working order and oil levels are correct. (5) Verify that the soil condition is good: moisture is less than 1.5% (1.0% preferred) by dry weight, oxygen is less than 2%, acidity is less than 0.5, and CO gas increases after heavy load seasons are not excessive. (6) Verify that the gauges are reading correctly when transformer loads are heavy. If correct field measurements differ from manufacturer's test report data, then investigate further before loading past nameplate criteria. (7) Verify with infrared camera or RTD during heavy load periods that the LTC top oil temperature relative to the main tank top oil temperature is correct. For normal LTC operation, the LTC top oil is cooler than the main tank top oil. A significant deviation from this indicates LTC abnormalities. (8) If the load current exceeds the bushing rating, do not exceed 110°C top oil temperature (IEEE, 1995). If bushing size is not known, perform an infrared scan of the bushing terminal during heavy load periods. Investigate further if the temperature of the top terminal cap is excessive. (9) Use winding power factor tests as a measure to confirm the integrity of a transformer's insulation system. This gives an indication of moisture and other contaminants in the system. High BIL transformers require low winding power factors (<0.5%), while low BIL transformers can tolerate higher winding power factors (<1.5%). (10) If the transformer is extremely dry (less than 0.5% by dry weight) and the load power factor is extremely good (0.99 lag to 0.99 lead), then add 10% to the above recommendations.

Source: Tillman, R. F., Jr, "Loading Power Transformers," in *The Electric Power Engineering Handbook*, L. L. Grigsby, Ed.: CRC Press, Boca Raton, FL, 2001.

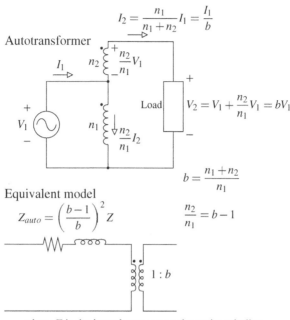

FIGURE 4.21
Autotransformer with an equivalent circuit.

4.9 Special Transformers

4.9.1 Autotransformers

An autotransformer is a winding on a core with a tap off the winding that provides voltage boost or buck. This is equivalent to a transformer with one winding in series with another (see Figure 4.21).

For small voltage changes, autotransformers are smaller and less costly than standard transformers. An autotransformer transfers much of the power directly through a wire connection. Most of the current passes through the lower-voltage series winding at the top, and considerably less current flows through the shunt winding.

Autotransformers have two main applications on distribution systems:

- *Voltage regulators* — A regulator is an autotransformer with adjustable taps that is normally capable of adjusting the voltage by ±10%.

- *Step banks* — Autotransformers are often used instead of traditional transformers on step banks and even substation transformers where the relative voltage change is moderate. This is normally voltage changes of less than a factor of three such as a 24.94Y/14.4 kV–12.47Y/7.2 kV bank.

The required rating of an autotransformer depends on the voltage change between the primary and secondary. The rating of each winding as a percentage of the load is

$$S = \frac{b-1}{b}$$

where
b = voltage change ratio, per unit

To obtain a 10% voltage change ($b = 1.1$), an autotransformer only has to be rated at 9% of the load kVA. For a 2:1 voltage change ($b = 2$), an autotransformer has to be rated at 50% of the load kVA. By comparison, a standard transformer must have a kVA rating equal to the load kVA.

The series impedance of autotransformers is less than an equivalent standard transformer. The equivalent series impedance of the autotransformer is

$$Z_{auto} = \left(\frac{b-1}{b}\right)^2 Z$$

where Z is the impedance across the entire winding. A 5%, 100-kVA conventional transformer has an impedance of 25.9 Ω at 7.2 kV line to ground. A 2:1 autotransformer ($b = 2$) with a load-carrying capability of 100 kVA and a winding rating of 50 kVA and also a 5% winding impedance has an impedance of 6.5 Ω, one-fourth that of a conventional transformer.

For three-phase applications on grounded systems, autotransformers are often connected in a grounded wye. Other possibilities are delta (each winding is phase to phase), open delta (same as a delta, but without one leg), and open wye. Because of the direct connection, it is not possible to provide ground isolation between the high- and low-voltage windings.

4.9.2 Grounding Transformers

Grounding transformers are sometimes used on distribution systems. A grounding transformer provides a source for zero-sequence current. Grounding transformers are sometimes used to convert a three-wire, ungrounded circuit into a four-wire, grounded circuit. Figure 4.22 shows the two most common grounding transformers. The zig-zag connection is the most widely used grounding transformer. Figure 4.23 shows how a grounding bank supplies current to a ground fault. Grounding transformers

Grounded Wye -- Delta   Zig-Zag Grounding Bank

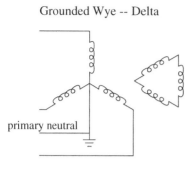

FIGURE 4.22
Grounding transformer connections.

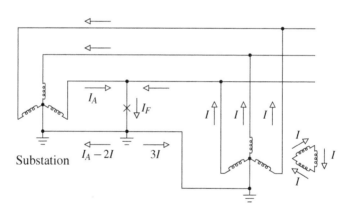

Sequence Equivalent

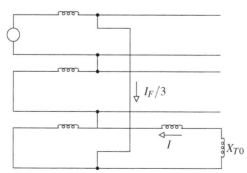

FIGURE 4.23
A grounding transformer feeding a ground fault.

used as the only ground source to a distribution circuit should be in service whenever the three-phase power source is in service. If the grounding transformer is lost, a line-to-ground causes high phase-to-neutral voltages on the unfaulted phases, and load unbalances can also cause neutral shifts and overvoltages.

A grounding transformer must handle the unbalanced load on the circuit as well as the duty during line-to-ground faults. If the circuit has minimal unbalance, then we can drastically reduce the rating of the transformer. It only has to be rated to carry short-duration (but high-magnitude) faults, normally a 10-sec or 1-min rating is used. We can also select the impedance of the grounding transformer to limit ground-fault currents.

Each leg of a grounding transformer carries one-third of the neutral current and has line-to-neutral voltage. So in a grounded wye – delta transformer, the total power rating including all three phases is the neutral current times the line-to-ground voltage:

$$S = V_{LG}I_N$$

A zig-zag transformer is more efficient than a grounded wye – delta transformer. In a zig-zag, each winding has less than the line-to-ground voltage, by a factor of $\sqrt{3}$, so the bank may be rated lower:

$$S = V_{LG}I_N/\sqrt{3}$$

ANSI/IEEE Std. 32-1972 requires a continuous rating of 3% for a 10-sec rated unit (which means the short-time rating is 33 times the continuous rating). A 1-min rated bank has a continuous current rating of 7%. On a 12.47-kV system supplying a ground-fault current of 6000 A, a zig-zag would need a 24.9-MVA rating. We will size the bank to handle the 24.9 MVA for 10 sec, which is equivalent to a 0.75-MVA continuous rating, so this bank could handle 180 A of neutral current continuously.

For both the zig-zag and the grounded wye – delta, the zero-sequence impedance equals the impedance between one transformer primary and its secondary.

Another application of grounding transformers is in cases of telephone interference due to current flow in the neutral/ground. By placing a grounding bank closer to the source of the neutral current, the grounding bank shifts some of the current from the neutral to the phase conductors to lower the neutral current that interferes with the telecommunication wires.

Grounding transformers are also used where utilities need a ground source during abnormal conditions. One such application is for a combination feeder that feeds secondary network loads and other non-network line-to-ground connected loads. If the network transformers are delta – grounded wye connected, the network will backfeed the circuit during a line-to-ground fault. If that happens while the main feeder breaker is open, the single-phase load on the unfaulted phases will see an overvoltage because the circuit is being back fed through the network loads as an ungrounded system. A grounding bank installed on the feeder prevents the overvoltage during backfeed conditions. Another similar application is found when applying distributed generators. A grounded wye – delta transformer is often specified as the interconnection transformer to prevent overvoltages if the generator drives an island that is separated from the utility source.

Even if a grounding bank is not the only ground source, it must be sized to carry the voltage unbalance. The zero-sequence current drawn by a bank is the zero-sequence voltage divided by the zero-sequence impedance:

$$I_0 = V_0/Z_0$$

Severe voltage unbalance can result when one phase voltage is opened upstream (usually from a blown fuse or a tripped single-phase recloser). In this case, the zero-sequence voltage equals the line-to-neutral voltage. The grounding bank will try to hold up the voltage on the opened phase and supply all of the load on that phase, which could severely overload the transformer.

4.10   Special Problems

4.10.1   Paralleling

Occasionally, crews must install distribution transformers, either at a changeover or for extra capacity. If a larger bank is being installed to replace an existing unit, paralleling the banks during the changeover eliminates the customer interruption. In order to parallel transformer banks, several criteria should be met:

- *Phasing* — The high and low-voltage connections must have the same phasing relationship. On three-phase units, banks of different connection types can be paralleled as long as they have compatible outputs: a delta – grounded wye may be paralleled with a grounded wye – grounded wye.
- *Polarity* — If the units have different polarity, they should be wired accordingly. (Flip one of the secondary connections.)
- *Voltage* — The phase-to-phase and phase-to-ground voltages on the outputs should be equal. Differences in turns ratios between the transformers will cause circulating current to flow through the transformers (continuously, even with zero load).

Before connecting the second transformer, crews should ensure that the secondary voltages are all zero or very close to zero (phase A to phase A, B to B, C to C, and the neutral to neutral).

If the percent impedances of the transformers are unequal, the load will not split in the same proportion between the two units. Note that this is the percent impedance, not the impedance in ohms. The unit with the lower percent impedance takes more of the current relative to its rating. For unequal impedances, the total bank must be derated (ABB, 1995) as

$$d = \frac{\dfrac{Z_2}{Z_1} K_1 + K_2}{K_1 + K_2}$$

where

K_1 = Capacity of the unit or bank with the *larger* percent impedance
K_2 = Capacity of the unit or bank with the *smaller* percent impedance
Z_1 = Percent impedance of unit or bank 1
Z_2 = Percent impedance of unit or bank 2

4.10.2 Ferroresonance

Ferroresonance is a special form of series resonance between the magnetizing reactance of a transformer and the system capacitance. A common form of ferroresonance occurs during single phasing of three-phase distribution transformers (Hopkinson, 1967). This most commonly happens on cable-fed transformers because of the high capacitance of the cables. The transformer connection is also critical for ferroresonance. An ungrounded primary connection (see Figure 4.24) leads to the highest magnitude ferroresonance. During single phasing (usually when line crews energize or deenergize the transformer with single-phase cutouts at the cable riser pole) a ferroresonant circuit between the cable capacitance and the transformer's magnetizing

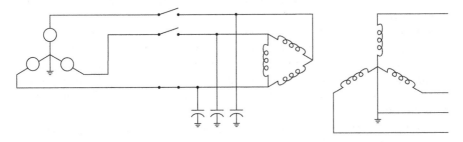

FIGURE 4.24
Ferroresonant circuit with a cable-fed transformer with an ungrounded high-side connection.

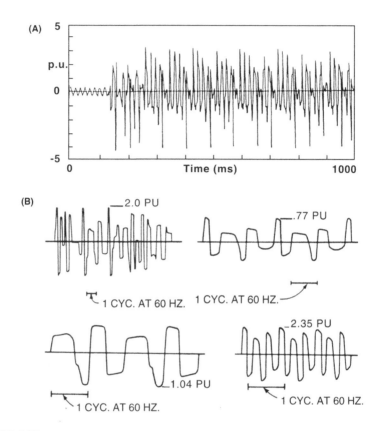

FIGURE 4.25

Examples of ferroresonance. (A) From Walling, R. A., Hartana, R. K., and Ros, W. J., "Self-Generated Overvoltages Due to Open-Phasing of Ungrounded-Wye Delta Transformer Banks," *IEEE Trans. Power Delivery*, 10(1), 526-533, January 1995. With permission. ©1995 IEEE. (B) Smith, D. R., Swanson, S. R., and Borst, J. D., "Overvoltages with Remotely-Switched Cable-Fed Grounded Wye-Wye Transformers," *IEEE Trans. Power Apparatus Sys.*, PAS-94(5), 1843-1853, 1975. With permission. ©1975 IEEE.

reactance drives voltages to as high as five per unit on the open legs of the transformer. The voltage waveform is normally distorted and often chaotic (see Figure 4.25).

Ferroresonance drove utilities to use three-phase transformer connections with a grounded-wye primary, especially on underground systems.

The chance of ferroresonance is determined by the capacitance (cable length) and by the core losses and other resistive load on the transformer (Walling et al., 1993). The core losses are an important part of the ferroresonant circuit.

Walling (1994) breaks down ferroresonance in a way that highlights several important aspects of this complicated phenomenon. Consider the simplified ferroresonant circuit in Figure 4.26. The transformer magnetizing branch has the core-loss resistance in parallel with a switched inductor. When the trans-

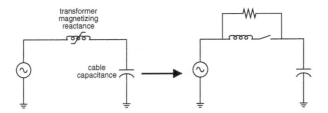

FIGURE 4.26
Simplified equivalent circuit of ferroresonance on a transformer with an ungrounded high-side connection.

former is unsaturated, the switched inductance is open, and the only connection between the capacitance and the system is through the core-loss resistance. When the core saturates, the capacitive charge dumps into the system (the switch in Figure 4.26 closes). The voltage overshoots and, as the core comes out of saturation, charge is again trapped on the capacitor (but of opposite polarity). This happens every half cycle (see Figure 4.27 for waveforms). If the core loss is large enough (or the resistive load on the transformer is large enough), the charge on the capacitor drains off before the next half cycle, and ferroresonance does not occur. The transformer core does not stay saturated long during each half cycle, just long enough to

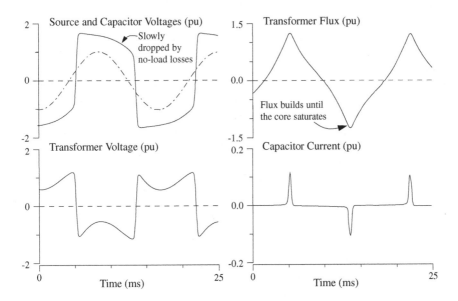

FIGURE 4.27
Voltages, currents and transformer flux during ferroresonance. (Adapted from Walling, R. A., "Ferroresonant Overvoltages in Today's Loss-Evaluated Distribution Transformers," IEEE/PES Transmission and Distribution Conference, 1994. With permission of the General Electric Company.)

TABLE 4.15

Transformer Primary Connections Susceptible to Ferroresonance

Susceptible Connections	Not Susceptible
Floating wye	Grounded wye made of three individual units
Delta	or units of triplex construction
Grounded wye with 3, 4, or 5-legged core	Open wye – open delta
construction	Line-to-ground connected single-phase units
Line-to-line connected single-phase units	

release the trapped charge on the capacitor. If the cable susceptance or even just the transformer susceptance is greater than the transformer core loss conductance, then ferroresonant overvoltages may occur.

In modern silicon-steel distribution transformers, the flux density at rated voltage is typically between 1.3 and 1.6 T. These operating flux densities slightly saturate the core (magnetic steel fully saturates at about 2 T). Because the core is operated near saturation, a small transient (such as switching) is enough to saturate the core. Once started, the ferroresonance self-sustains. The resonance repeatedly saturates the transformer every half cycle.

Table 4.15 shows what types of transformer connections are susceptible to ferroresonance. To avoid ferroresonance on floating wye – delta transformers, some utilities temporarily ground the wye on the primary side of floating wye – delta connections during switching operations.

Ferroresonance can occur on transformers with a grounded primary connection if the windings are on a common core such as the five-legged core transformer [the magnetic coupling between phases completes the ferroresonant circuit (Smith et al., 1975)]. The five-legged core transformer connected as a grounded wye – grounded wye is the most common underground transformer configuration. Ferroresonant overvoltages involving five-legged core transformers normally do not exceed two per unit.

Ferroresonance is a function of the cable capacitance and the transformer no-load losses. The lower the losses relative to the capacitance, the higher the ferroresonant overvoltage can be. For transformer configurations that are susceptible to ferroresonance, ferroresonance can occur approximately when

$$B_C \geq P_{NL}$$

where
    B_C = capacitive reactive power per phase, vars
    P_{NL} = core loss per phase, W

The capacitive reactive power on one phase in vars depends on the voltage and the capacitance as

$$B_C = \frac{V_{kV}^2}{3} 2\pi f C$$

where
V_{kV} = rated line-to-line voltage, kV
f = frequency, Hz
C = capacitance from one phase to ground, μF

Normally, ferroresonance occurs without equipment failure if the crew finishes the switching operation in a timely manner. The loud banging, rumbling, and rattling of the transformer during ferroresonance may alarm line crews. Occasionally, ferroresonance is severe enough to fail a transformer. The overvoltage stresses the transformer insulation, and the repeated saturation may cause tank heating as flux leaves the core (although many modes of ferroresonance barely saturate the transformer and do not cause significant tank heating). Surge arresters are the most likely equipment casualty. In attempting to limit the ferroresonant overvoltage, an arrester may absorb more current than it can handle and thermally run away. Gapped silicon-carbide arresters were particularly prone to failure, as the gap could not reseal the repeated sparkovers from a long-duration overvoltage. Gapless metal-oxide arresters are much more resistant to failure from ferroresonance and help hold down the overvoltages. Ferroresonant overvoltages may also fail customer's equipment from high secondary voltages. Small end-use arresters are particularly susceptible to damage.

Ferroresonance is more likely with

- *Unloaded transformers* — Ferroresonance disappears with load as little as a few percent of the transformer rating.
- *Higher primary voltages* — Shorter cable lengths are required for ferroresonance. Resonance is more likely even without cables, just due to the internal capacitance of the transformer. With higher voltages, the capacitances do not change significantly (cable capacitance increases just slightly because of thicker insulation), but vars are much higher for the same capacitance.
- *Smaller transformers* — Smaller no-load losses.
- *Low-loss transformers* — Smaller no-load losses.

Severe ferroresonance with voltages reaching peaks of 4 or 5 per unit occurs on three-phase transformers with an ungrounded high-voltage winding during single-pole switching. If the transformer is fed by underground cables and crews switch the transformer remotely, ferroresonance is likely.

On overhead circuits, ferroresonance is common with ungrounded primary connections on 25- and 35-kV distribution systems. At these voltages, the internal capacitance of most transformers is enough to ferroresonate. The use of low-loss transformers has caused ferroresonance to appear on overhead 15-kV distribution systems as well. Amorphous core and low-loss silicon-steel core transformers have much lower core losses than previous designs. With less core losses, ferroresonance happens with lower amounts of capacitance. Tests by the Southern California Edison Company on three-phase transformers with ungrounded primary connections found that ferroresonance occurred when the capacitive power per phase exceeded the transformer's no-load losses per phase by the following relationship (Jufer, 1994):

$$B_C \geq 1.27 P_{NL}$$

The phase-to-ground capacitance of overhead transformers is primarily due to the capacitance between the primary and secondary windings (the secondary windings are almost at zero potential). A typical 25-kVA transformer has a phase-to-ground capacitance of about 2 nF (Walling et al., 1995). For a 7.2-kV line-to-ground voltage, 0.002 µF is 39 vars. So, if the no-load losses are less than 39 vars/1.27 = 30.7 W per phase, the transformer may ferroresonate under single-pole switching.

Normally, ferroresonance occurs on three-phase transformers, but ferroresonance can occur on single-phase transformers if they are connected phase to phase, and one of the phases is opened either remotely or at the transformer. Jufer (1994) found that small single-phase padmounted transformers connected phase to phase ferroresonate when remotely switched with relatively short cable lengths. Their tests of silicon-steel core transformers found that a 25-kVA transformer resonated with 50 ft (15 m) of 1/0 XLPE cable at 12 kV. A 50-kVA transformer resonated with 100 ft of cable, and a 75-kVA unit resonated with 150 ft of the cable. Peak primary voltages reached 3 to 4 per unit. Secondary-side peaks were all under 2 per unit. Longer cables produced slightly higher voltages during ferroresonance. Jufer found that ferroresonance didn't occur if the resistive load in watts per phase (including the transformer's no-load losses and the resistive load on the secondary) exceeded 1.15 times the capacitive vars per phase ($P_{NL} + P_L > 1.15B_C$). Bohmann et al. (1991) describes a feeder where single-phase loads were switched to a phase-to-phase configuration, and the reconfiguration caused a higher-than-normal arrester failure rate that was attributed to ferroresonant conditions on the circuit.

It is widely believed that a grounded-wye primary connection eliminates ferroresonance. This is not true if the three-phase transformer has windings on a common core. The most common underground three-phase distribution transformer has a five-legged wound core. The common core couples the phases. With the center phase energized and the outer phases open, the coupling induces 50% voltage in the outer phases. Any load on the outer two phases is effectively in series with the voltage induced on the center phase. Because the coupling is indirect and the open phase capacitance is in parallel with a transformer winding to ground, this type of ferroresonance is not as severe as ferroresonance on configurations with an ungrounded primary winding. Overvoltages rarely exceed 2.5 per unit.

Five-legged core ferroresonance also depends on the core losses of the transformer and the phase-to-ground capacitance. If the capacitive vars exceed the resistive load in watts, ferroresonance may occur. Higher capac-

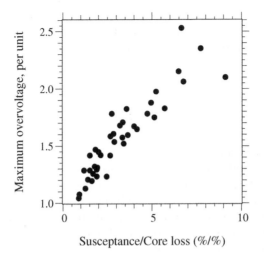

FIGURE 4.28
Five-legged core ferroresonance as a function of no-load losses and line-to-ground capacitance. (Adapted from Walling, R. A., Barker, K. D., Compton, T. M., and Zimmerman, L. E., "Ferroresonant Overvoltages in Grounded Wye-Wye Padmount Transformers with Low-Loss Silicon Steel Cores," *IEEE Trans. Power Delivery*, 8(3), 1647-60, July 1993. With permission. ©1993 IEEE.)

itances — longer cable lengths — generally cause higher voltages (see Figure 4.28). To limit peak voltages to below 1.25 per unit, the capacitive power must be limited such that [equivalent to that proposed by Walling (1992)]:

$$B_C \leq 1.86 P_{NL}$$

with B_C in vars and P_{NL} in watts; both are per phase.

Ferroresonance can occur with five-legged core transformers, even when switching at the transformer terminals, due to the transformer's internal line-to-ground capacitance. On 34.5-kV systems, transformers smaller than 500 kVA may ferroresonate if single-pole switched right at the transformer terminals. Even on 15-kV class systems where crews can safely switch all but the smallest 5-legged core transformers at the terminals, we should include the transformer's capacitance in any cable length calculation; the transformer's capacitance is equivalent to several feet (meters) of cable. The capacitance from line-to-ground is mainly due to the capacitance between the small paper-filled layers of the high-voltage winding. This capacitance is very difficult to measure since it is in parallel with the coil. Walling (1992) derived an empirical equation to estimate the line-to-ground transformer capacitance per phase in μF:

$$C = \frac{0.000469 S_{kVA}^{0.4}}{V_{kV}^{0.25}}$$

where
S_{kVA} = transformer three-phase kVA rating
V_{kV} = rated line-to-line voltage in kV

In vars, this is

$$B_C = 0.000982 f V_{kV}^{1.75} S_{kVA}^{0.4}$$

where f is the system frequency, Hz.

To determine whether the transformer no-load losses exceed the capacitive power, the transformer's datasheet data is most accurate. For coming up with generalized guidelines, using such data is not realistic since so many different transformer makes and models are ordered. Walling (1992) offered the following approximation between the three-phase transformer rating and the no-load losses in watts per phase:

$$P_{NL} = S_{kVA} \left(4.54 - 1.13 \log_{10} \left(S_{kVA} \right) \right) / 3$$

Walling (1992) used his approximations of transformer no-load losses and transformer capacitance to find cable length criteria for remote single-pole switching. Consider a 75-kVA 3-phase 5-legged core transformer at 12.47 kV. Using these approximations, the no-load losses are 60.5 W per phase, and the transformer's capacitance is 27.4 vars per phase. To keep the voltage under 1.25 per unit, the total vars allowed per phase is 1.86(60.5W) = 111.9 vars. So, the cable can add another 84.5 vars before we exceed the limit. At 12.47 kV, a 4/0 175-mil XLPE cable has a capacitance of 0.412 µF/mi, which is 1.52 vars per foot. For this cable, 56 ft is the maximum length that we should switch remotely. Beyond that, we may have ferroresonance above 1.25 per unit. Table 4.16 shows similar criteria for several three-phase transformers and voltages. The table shows critical lengths for 4/0 cables; smaller cables have less capacitance, so somewhat longer lengths are permissible. At 34.5 kV, crews should only remotely switch larger banks.

Another situation that can cause ferroresonance is when a secondary has ungrounded power factor correction capacitors. Resonance can even occur on a grounded wye – grounded wye connection with three separate transformers. With one phase open on the utility side, the ungrounded capacitor bank forms a series resonance with the magnetizing reactance of the open leg of the grounded-wye transformer.

Ferroresonance most commonly happens when switching an unloaded transformer. It also usually happens with manual switching; ferroresonance can occur because a fault clears a single-phase protective device, but this is much less common. The main reason that ferroresonance is unlikely for most situations using a single-phase protective device is that either the fault or the existing load on the transformer prevents ferroresonance.

TABLE 4.16

Cable Length Limits in Feet for Remote Single-Pole
Switching to Limit Ferroresonant Overvoltages to
Less than 1.25 per Unit

	Critical Cable Lengths, ft		
Transformer Rating kVA	12.47 kV 4/0 XLPE 175 mil 0.412 µF/mi 1.52 vars/ft	24.94 kV 4/0 XLPE 260 mil 0.261 µF/mi 4.52 vars/ft	34.5 kV 4/0 XLPE 345 mil 0.261 µF/mi 7.08 vars/ft
75	56	5	0
112.5	81	10	0
150	103	16	0
225	144	26	1
300	181	36	6
500	265	59	16
750	349	82	27
1000	417	100	36
1500	520	128	49
2000	592	146	56

If the fuse is a tap fuse and several customers are on a section, the transformers will have somewhat different characteristics, which lowers the probability of ferroresonance (and ferroresonance is less likely with larger transformers).

Solutions to ferroresonance include

- Using a higher-loss transformer
- Using a three-phase switching device instead of a single-phase device
- Switching right at the transformer rather than at the riser pole
- Using a transformer connection not susceptible to ferroresonance
- Limiting remote switching of transformers to cases where the capacitive vars of the cable are less than the transformer's no load losses

Arrester application on transformer connections susceptible to ferroresonance brings up several interesting points. Ferroresonance can slowly heat arresters until failure. Ferroresonance is a weak source; even though the per-unit magnitudes are high, the voltage collapses when the arrester starts to conduct (we cannot use the arresters time-overvoltage curve [TOV] to predict failure). Normally, extended ferroresonance of several minutes can occur before arresters are heated enough to enter thermal runaway. The most vulnerable arresters are those that are tightly applied relative to the voltage rating. Tests by the DSTAR group for ferroresonance on 5-legged core transformers in a grounded wye – grounded wye connection (Lunsford, 1994; Walling et al., 1994) found

- Arrester currents were always less than 2 A.
- Under-oil arresters, which have superior thermal characteristics, reached thermal stability and did not fail.
- Porcelain-housed arresters showed slow heating — sometimes enough to fail, sometimes not, depending on the transformer type, cable lengths, and arrester type. Elbow arresters showed slow heating — slower than the riser-pole arresters. Failure times for either type were typically longer than 30 min.

With normal switching times of less than one minute, arresters do not have enough time to heat and fail. Crews should be able to safely switch transformers under most circumstances. Load — even 5% of the transformer rating — prevents ferroresonance in most cases. The most danger is with unloaded transformers. If an arrester fails, the failure may not operate the disconnect, which can lead to a dangerous scenario. When a line worker recloses the switch, the stiff power-frequency source will fail the arrester. The disconnect should operate and draw an arc. On occasion, the arrester may violently shatter.

One option to limit the exposure of the arresters is to put the arresters upstream of the switch. At a cable riser pole this is very difficult to do without seriously compromising the lead length of the arrester.

4.10.3   Switching Floating Wye – Delta Banks

Floating wye – delta banks present special concerns. As well as being prone to ferroresonance, single-pole switching can cause overvoltages due to a neutral shift. On a floating wye – delta, the secondary delta connection fixes the transformer's primary neutral close to ground potential. After one phase of the primary wye is opened, the neutral can float far from ground. This causes overvoltages, both on the secondary side and the primary side. The severity depends on the balance of the load.

When crews open one of the power-leg phases, if there is no three-phase load and only the single-phase load on the lighting leg of the transformer, the open primary voltage V_{open} reaches 2.65 times normal as shown in Figure 4.29. The voltage across the open switch also sees high voltage. The voltage from B to B′ in Figure 4.29 can reach over 2.75 per unit. Secondary line-to-line voltages on the power legs can reach 1.73 per unit. The secondary delta forces the sum of the three primary line-to-neutral voltages to be equal. With single-phase load on phase C and no other load, the neutral shifts to the C-phase voltage. The delta winding forces $V_{B'N}$ to be equal to $-V_{AN}$, significantly shifting the potential of point B′.

The line-to-ground voltage on the primary-side of the transformer on the open phase is a function of the load unbalance on the secondary. Given the ratio of the single-phase load to the three-phase load, this voltage is [assuming

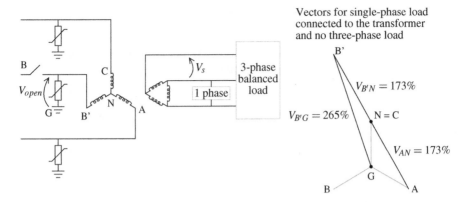

FIGURE 4.29
Neutral-shift overvoltages on a floating wye – delta transformer during single-pole switching.

passive loads and that the power factor of the three-phase load equals that of the single-phase load (Walling et al., 1995)]

$$V_{open} = \frac{\sqrt{7K^2 + K + 1}}{K + 2}$$

where

$$K = \frac{\text{Single-phase load}}{\text{Balanced three-phase load}}$$

On the secondary side, the worst of the two line-to-line voltages across the power legs have the following overvoltages depending on loading balance (PTI, 1999):

$$V_s = \sqrt{3}\,\frac{K + 1}{K + 2}$$

Figure 4.30 shows these voltages as a function of the ratio K.

Contrary to a widespread belief, transformer saturation does not significantly reduce the overvoltage. Walling et al.'s (1995) EMTP simulations showed that saturation did not significantly reduce the peak voltage magnitude. Saturation does distort the waveforms significantly and reduces the energy into a primary arrester.

Some ways to avoid these problems are

- *Use another connection* — The best way to avoid problems with this connection is to use some other connection. Some utilities do not

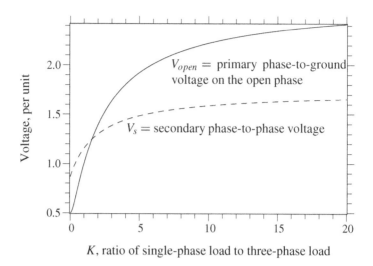

FIGURE 4.30
Neutral-shift overvoltages as a function of the load unbalance.

offer an open wye – delta connection and instead move customers to grounded-wye connections.

- *Neutral grounding* — Ground the primary-wye neutral during switching operations, either with a temporary grounding jumper or install a cutout. This prevents the neutral-shift and ferroresonant overvoltage. The ground-source effects during the short-time switching are not a problem. The line crew must remove the neutral jumper after switching. Extended operation as a grounding bank can overheat the transformer and interfere with a circuit's ground-fault protection schemes.

- *Switching order* — Neutral shifts (but not ferroresonance) are eliminated by always switching in the lighting leg last and taking it out first.

Arrester placement is a sticky situation. If the arrester is upstream of the switch, it does not see the neutral-shift/ferroresonant overvoltage. But the transformer is not protected against the overvoltages. Arresters downstream of the switch protect the transformer but may fail. One would rather have an arrester failure than a transformer failure, unless the failure is near a line crew (since an arrester is smaller, it is more likely than a transformer to explode violently — especially porcelain-housed arresters). Another concern was reported by Walling (2000): during switching operations, 10-per-unit overvoltage bursts for 1/4 cycle ringing at about 2 kHz when closing in the second phase. These were found in measurements during full-scale tests and also in simulations. This transient repeats every cycle with a declining peak magnitude for more than one second. If arresters are downstream from the switches, they can easily control the overvoltage. But if they are upstream of the switches, this high voltage stresses the transformer insulation.

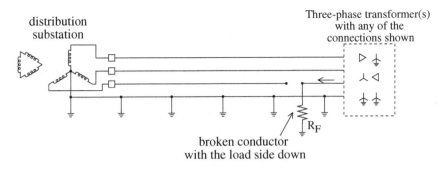

FIGURE 4.31
Backfeed to a downed conductor.

Overall, grounding the transformer's primary neutral is the safest approach.

4.10.4 Backfeeds

During a line-to-ground fault where a single-phase device opens, current may backfeed through a three-phase load (see Figure 4.31). It is a common misconception that this type of backfeed can only happen with an ungrounded transformer connection. Backfeed can also occur with a grounded three-phase connection. This creates hazards to the public in downed wire situations. Even though it is a weak source, the backfed voltage is just as dangerous. Lineworkers also have to be careful. A few have been killed after touching wires downstream of open cutouts that they thought were deenergized.

The general equations for the backfeed voltage and current based on the sequence impedances of the load (Smith, 1994) are

$$I_F = \frac{(A - 3Z_0 Z_2) V}{3Z_0 Z_1 Z_2 + R_F A}$$

$$V_F = R_F I_F$$

where
$A = Z_0 Z_1 + Z_1 Z_2 + Z_0 Z_2$
Z_1 = positive-sequence impedance of the load, Ω
Z_2 = negative-sequence impedance of the load, Ω
Z_0 = zero-sequence impedance of the load, Ω
R_F = fault resistance, Ω
V = line-to-neutral voltage, V

The line and source impedances are left out of the equations because they are small relative to the load impedances. Under an open circuit with no fault ($R_F = \infty$), the backfeed voltage is

$$V_F = \frac{(A - 3Z_0 Z_2) V}{A}$$

For an ungrounded transformer connection ($Z_0 = \infty$), the backfeed current is

$$I_F = \frac{(Z_1 - 2Z_2) V}{3Z_1 Z_2 + R_F (Z_1 + Z_2)}$$

The backfeed differs depending on the transformer connection and the load:

- Grounded wye – grounded wye transformer connection
  - Will not backfeed the fault when the transformer is unloaded or has balanced line-to-ground loads (no motors). It will backfeed the fault with line-to-line connected load (especially motors).
- Ungrounded primary transformer
  - Will backfeed the fault under no load. It may not be able to provide much current with no load, but there can be significant voltage on the conductor. Motor load will increase the backfeed current available.

Whether it is a grounded or ungrounded transformer, the available backfeed current depends primarily on the connected motor load. Motors dominate since they have much lower negative-sequence impedance; typically it is equal to the locked-rotor impedance or about 15 to 20%. With no fault impedance ($R_F = 0$), the backfeed current is approximately:

$$I_F = \frac{M_{kVA}}{9V_{LG,kV} \cdot Z_{2,pu}}$$

where M_{kVA} is the three-phase motor power rating in kVA (and we can make the common assumption that 1 hp = 1 kVA), $V_{LG,kV}$ is the line-to-ground voltage in kV, and $Z_{2,pu}$ is the per-unit negative-sequence (or locked-rotor) impedance of the motor(s). Figure 4.32 shows the variation in backfeed current versus motor kVA on the transformer for a 12.47-kV system (assuming $Z_{2,pu} = 0.15$).

The voltage on the open phases depends on the type of transformer connection and the portion of the load that is motors. Figure 4.33 shows the backfeed voltage for an open circuit and for a typical high-impedance fault ($R_F = 200\ \Omega$).

As discussed in Chapter 7, the maximum sustainable arc length in inches is roughly $l = \sqrt{I} \cdot V$ where I is the rms current in amperes, and V is the

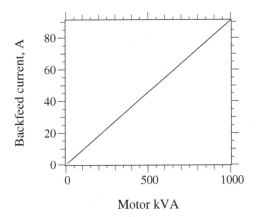

FIGURE 4.32
Available backfeed current on a 12.47-kV circuit (grounded wye – grounded wye or an ungrounded connection, $R_F = 0$).

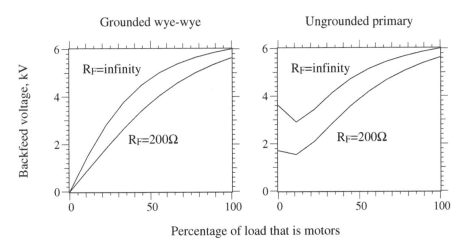

FIGURE 4.33
Available backfeed voltage on a 12.47-kV circuit.

voltage in kV. For a line-to-ground fault on a 12.47-kV circuit, if the backfeed voltage is 4 kV with 50 A available (typical values from Figure 4.32 and Figure 4.33), the maximum arc length is 28 in. (0.7 m). Even though the backfeed source is weak relative to a traditional fault source, it is still strong enough to maintain a significant arc during backfeeds.

In summary, the backfeed voltage is enough to be a safety hazard to workers or the public (e.g., in a wire down situation). The available backfeed is a stiff enough source to maintain an arc of significant length. The arc can continue to cause damage at the fault location during a backfeed condition. It may also spark and sputter at a low level. Options to reduce the chances of backfeed problems include:

- Make sure crews follow safety procedures (if it is not grounded, it is not dead).
- Follow standard practices regarding downed conductors including proper line designs and maintenance, public education, and worker training.

Another option is to avoid single-pole protective devices (switches, fuses, or single-phase reclosers) upstream of three-phase transformer banks. Most utilities have found that backfeeding problems are not severe enough to warrant not using single-pole protective devices.

To analyze more complicated arrangements, use a steady-state circuit analysis program (EMTP has this capability). Most distribution fault analysis programs cannot handle this type of complex arrangement.

4.10.5 Inrush

When a transformer is first energized or reenergized after a short interruption, the transformer may draw *inrush* current from the system due to the core magnetization being out of sync with the voltage. The inrush current may approach short-circuit levels, as much as 40 times the transformer's full-load current. Inrush may cause fuses, reclosers, or relays to falsely operate. It may also falsely operate faulted-circuit indicators or cause sectionalizers to misoperate.

When the transformer is switched in, if the system voltage and the transformer core magnetization are not in sync, a magnetic transient occurs. The transient drives the core into saturation and draws a large amount of current into the transformer.

The worst inrush occurs with residual flux left on the transformer core. Consider Figure 4.34 and Figure 4.35, which shows the worst-case scenario. A transformer is deenergized near the peak core flux density (B_{max}), when the voltage is near zero. The flux decays to about 70% of the maximum and holds there (the residual flux, B_r). Some time later, the transformer is reenergized at a point in time when the flux would have been at its negative peak; the system voltage is crossing through zero and rising positively. The positive voltage creates positive flux that adds to the residual flux already on the transformer core (remember, flux is the time integral of the voltage). This quickly saturates the core; the effective magnetizing branch drops to the air-core impedance of the transformer.

The air core impedance is roughly the same magnitude as the transformer's leakage impedance. Flux controls the effective impedance, so when the core saturates, the small impedance pulls high-magnitude current from the system. The core saturates in one direction, so the transformer draws pulses of inrush every other half cycle with a heavy dc component. The dc offset introduced by the switching decays away relatively quickly.

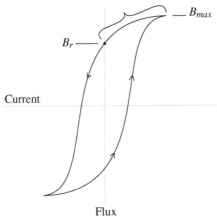

FIGURE 4.34
Hysteresis curve showing the residual flux during a circuit interruption.

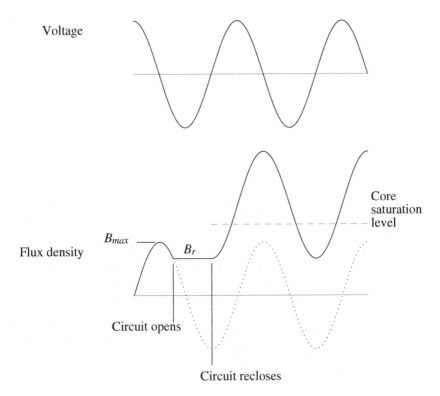

FIGURE 4.35
Voltage and flux during worst-case inrush.

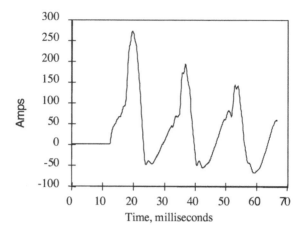

FIGURE 4.36
Example inrush current measured at a substation (many distribution transformers together). (Copyright © 1996. Electric Power Research Institute. TR-106294-V3. *An Assessment of Distribution System Power Quality: Volume 3: Library of Distribution System Power Quality Monitoring Case Studies.* Reprinted with permission.)

Figure 4.36 shows an example of inrush following a reclose operation measured at the distribution substation breaker.

Several factors significantly impact inrush:

- *Closing point* — The point where the circuit closes back in determines how close the core flux can get to its theoretical maximum. The worst case is when the flux is near its peak. Fortunately, this is also when the voltage is near zero, and switches tend to engage closer to a voltage peak (an arc tends to jump the gap).

- *Design flux* — A transformer that is designed to operate lower on the saturation curve draws less inrush. Because there is more margin between the saturation point and the normal operating region, the extra flux during switching is less likely to push the core into saturation.

- *Transformer size* — Larger transformers draw more inrush. Their saturated impedances are smaller. But, on a per-unit basis relative to their full-load capability, smaller transformers draw more inrush. The inrush into smaller transformers dies out more quickly.

- *Source impedance* — Higher source impedance relative to the transformer size limits the current that the transformer can pull from the system. The peak inrush with significant source impedance (Westinghouse Electric Corporation, 1950) is

$$i_{peak} = \frac{i_0}{1 + i_0 X}$$

where
> i_0 = peak inrush without source impedance in per unit of the transformer rated current
>
> X = source impedance in per unit on the transformer kVA base

Other factors have less significance. The load on the transformer does not significantly change the inrush. For most typical loading conditions, the current into the transformer will interrupt at points that still leave about 70% of the peak flux on the core.

While interruptions generally cause the most severe inrush, other voltage disturbances may cause inrush into a transformer. Voltage transients and especially voltage with a dc component can saturate the transformer and cause inrush. Some examples are:

- *Voltage sags* — Upon recovery from a voltage sag from a nearby fault, the sudden rise in voltage can drive a transformer into saturation.

- *Sympathetic inrush* — Energizing a transformer can cause a nearby transformer to also draw inrush. The inrush into the switched transformer has a significant dc component that causes a dc voltage drop. The dc voltage can push the other transformer into saturation and draw inrush.

- *Lightning* — A flash to the line near the transformer can push the transformer into saturation.

References

ABB, *Distribution Transformer Guide*, 1995.

Alexander Publications, *Distribution Transformer Handbook*, 2001.

ANSI C57.12.40-1982, *American National Standard Requirements for Secondary Network Transformers, Subway and Vault Types (Liquid Immersed)*.

ANSI/IEEE C57.12.24-1988, *American National Standard Underground-type Three-Phase Distribution Transformers, 2500 kVA and Smaller; High Voltage 34 500 GrdY/19 200 V and Below; Low Voltage 480 V and Below — Requirements*.

ANSI/IEEE C57.12.80-1978, *IEEE Standard Terminology for Power and Distribution Transformers*.

ANSI/IEEE C57.91-1981, *IEEE Guide for Loading Mineral-Oil-Immersed Overhead and Pad-Mounted Distribution Transformers Rated 500 kVA and Less with 65 Degrees C Or 55 Degrees C Average Winding Rise*.

ANSI/IEEE C57.105-1978, *IEEE Guide for Application of Transformer Connections in Three-Phase Distribution Systems*.

ANSI/IEEE Std. 32-1972, *IEEE Standard Requirements, Terminology, and Test Procedure for Neutral Grounding Devices*.

Blume, L. F., Boyajian, A., Camilli, G., Lennox, T. C., Minneci, S., and Montsinger, V. M., *Transformer Engineering*, Wiley, New York, 1951.

Bohmann, L. J., McDaniel, J., and Stanek, E. K., "Lightning Arrester Failures and Ferroresonance on a Distribution System," IEEE Rural Electric Power Conference, 1991.

CEA 485 T 1049, *On-line Condition Monitoring of Substation Power Equipment Utility Needs*, Canadian Electrical Association, 1996.

CIGRE working group 12.05, "An International Survey on Failure in Large Power Transformer Service," *Electra*, no. 88, pp. 21–48, 1983.

EEI, "A Method for Economic Evaluation of Distribution Transformers," March, 28–31, 1981.

EPRI TR-106294-V3, *An Assessment of Distribution System Power Quality: Volume 3: Library of Distribution System Power Quality Monitoring Case Studies*, Electric Power Research Institute, Palo Alto, CA, 1996.

Gangel, M. W. and Propst, R. F., "Distribution Transformer Load Characteristics," *IEEE Transactions on Power Apparatus and Systems*, vol. 84, pp. 671–84, August 1965.

Grainger, J. J. and Kendrew, T. J., "Evaluation of Technical Losses on Electric Distribution Systems," CIRED, 1989.

Hopkinson, F. H., "Approximate Distribution Transformer Impedances," General Electric Internal Memorandum, 1976. As cited by Kersting, W. H. and Phillips, W. H., "Modeling and Analysis of Unsymmetrical Transformer Banks Serving Unbalanced Loads," Rural Electric Power Conference, 1995.

Hopkinson, R. H., "Ferroresonant Overvoltage Control Based on TNA Tests on Three-Phase Delta-Wye Transformer Banks," *IEEE Transactions on Power Apparatus and Systems*, vol. 86, pp. 1258–65, October 1967.

IEEE C57.12.00-2000, *IEEE Standard General Requirements for Liquid-Immersed Distribution, Power, and Regulating Transformers*.

IEEE Std. 493-1997, *IEEE Recommended Practice for the Design of Reliable Industrial and Commercial Power Systems (Gold Book)*.

IEEE Std. C57.91-1995, *IEEE Guide for Loading Mineral-Oil-Immersed Transformers*.

IEEE Task Force Report, "Secondary (Low-Side) Surges in Distribution Transformers," *IEEE Transactions on Power Delivery*, vol. 7, no. 2, pp. 746–56, April 1992.

Jufer, N. W., "Southern California Edison Co. Ferroresonance Testing of Distribution Transformers," IEEE/PES Transmission and Distribution Conference, 1994.

Long, L. W., "Transformer Connections in Three-Phase Distribution Systems," in Power Transformer Considerations of Current Interest to the Utility Engineer, 1984. IEEE Tutorial Course, 84 EHO 209-7-PWR.

Lunsford, J., "MOV Arrester Performance During the Presence of Ferroresonant Voltages," IEEE/PES Transmission and Distribution Conference, 1994.

Nickel, D. L., "Distribution Transformer Loss Evaluation. I. Proposed Techniques," *IEEE Transactions on Power Apparatus and Systems*, vol. PAS-100, no. 2, pp. 788–97, February 1981.

NRECA RER Project 90-8, *Underground Distribution System Design and Installation Guide*, National Rural Electric Cooperative Association, 1993.

ORNL-6804/R1, *The Feasibility of Replacing or Upgrading Utility Distribution Transformers During Routine Maintenance*, Oak Ridge National Laboratory, U.S. Department of Energy, 1995.

ORNL-6847, *Determination Analysis of Energy Conservation Standards for Distribution Transformers*, Oak Ridge National Laboratory, U.S. Department of Energy, 1996.

ORNL-6925, *Supplement to the "Determination Analysis" (ORNL-6847) and Analysis of the NEMA Efficiency Standard for Distribution Transformers*, Oak Ridge National Laboratory, U.S. Department of Energy, 1997.

ORNL-6927, *Economic Analysis of Efficient Distribution Transformer Trends*, Oak Ridge National Laboratory, U.S. Department of Energy, 1998.

PTI, "Distribution Transformer Application Course Notes," Power Technologies, Inc., Schenectady, NY, 1999.

Rusch, R. J. and Good, M. L., "Wyes and Wye Nots of Three-Phase Distribution Transformer Connections," IEEE Rural Electric Power Conference, 1989.

Sankaran, C., "Transformers," in *The Electrical Engineering Handbook*, R. C. Dorf, Ed.: CRC Press, Boca Raton, FL, 2000.

Seevers, O. C., *Management of Transmission & Distribution Systems*, PennWell Publishing Company, Tulsa, OK, 1995.

Smith, D. R., "Impact of Distribution Transformer Connections on Feeder Protection Issues," Texas A&M Annual Conference for Protective Relay Engineers, March 1994.

Smith, D. R., Braunstein, H. R., and Borst, J. D., "Voltage Unbalance in 3- and 4-Wire Delta Secondary Systems," *IEEE Transactions on Power Delivery*, vol. 3, no. 2, pp. 733–41, April 1988.

Smith, D. R., Swanson, S. R., and Borst, J. D., "Overvoltages with Remotely-Switched Cable-Fed Grounded Wye-Wye Transformers," *IEEE Transactions on Power Apparatus and Systems*, vol. PAS-94, no. 5, pp. 1843–53, 1975.

Tillman, R. F., Jr, "Loading Power Transformers," in *The Electric Power Engineering Handbook*, L. L. Grigsby, Ed.: CRC Press, Boca Raton, FL, 2001.

Walling, R. A., "Ferroresonance Guidelines for Modern Transformer Applications," in Final Report to the Distribution Systems Testing, Application, and Research (DSTAR) Consortium: General Electric, Industrial and Power Systems, Power Systems Engineering Department, 1992. As cited in NRECA RER Project 90-8, 1993.

Walling, R. A., "Ferroresonant Overvoltages in Today's Loss-Evaluated Distribution Transformers," IEEE/PES Transmission and Distribution Conference, 1994.

Walling, R. A., 2000. Verbal report at the fall IEEE Surge Protective Devices Committee Meeting.

Walling, R. A., Barker, K. D., Compton, T. M., and Zimmerman, L. E., "Ferroresonant Overvoltages in Grounded Wye-Wye Padmount Transformers with Low-Loss Silicon Steel Cores," *IEEE Transactions on Power Delivery*, vol. 8, no. 3, pp. 1647–60, July 1993.

Walling, R. A., Hartana, R. K., Reckard, R. M., Sampat, M. P., and Balgie, T. R., "Performance of Metal-Oxide Arresters Exposed to Ferroresonance in Padmount Transformers," *IEEE Transactions on Power Delivery*, vol. 9, no. 2, pp. 788–95, April 1994.

Walling, R. A., Hartana, R. K., and Ros, W. J., "Self-Generated Overvoltages Due to Open-Phasing of Ungrounded-Wye Delta Transformer Banks," *IEEE Transactions on Power Delivery*, vol. 10, no. 1, pp. 526–33, January 1995.

Westinghouse Electric Corporation, *Electrical Transmission and Distribution Reference Book*, 1950.

All hell broke loose, we had a ball of fire that went phase to phase shooting fire out the xfmer vents like a flame thrower showering slag on the linemen and sent the monster galloping down the line doing the Jacobs ladder effect for 2 spans before it broke ...

The next time you're closing in on that new shiny xfmer out of the shop, think about the night we got a lemon.

anonymous poster
www.powerlineman.com

5

Voltage Regulation

One of a utility's core responsibilities is to deliver voltage to customers within a suitable range, so utilities must regulate the voltage. On distribution circuits, voltage drops due to current flowing through the line impedances. Primary and secondary voltage drop can be allocated as necessary along the circuit to provide end users with suitable voltage. Voltage regulators — in the substation or on feeders — can adjust primary voltage. This chapter discusses voltage regulators and regulation standards and techniques.

5.1 Voltage Standards

Most regulatory bodies and most utilities in America follow the ANSI voltage standards (ANSI C84.1-1995). This standard specifies acceptable operational ranges at two locations on electric power systems:

- *Service voltage* — The service voltage is the point where the electrical systems of the supplier and the user are interconnected. This is normally at the meter. Maintaining acceptable voltage at the service entrance is the *utility's* responsibility.

- *Utilization voltage* — The voltage at the line terminals of utilization equipment. This voltage is the *facility's* responsibility. Equipment manufacturers should design equipment which operates satisfactorily within the given limits.

The standard allows for some voltage drop within a facility, so service voltage requirements are tighter than utilization requirements.

The standard also defines two ranges of voltage:

- *Range A* — Most service voltages are within these limits, and utilities should design electric systems to provide service voltages within

TABLE 5.1

ANSI C84.1 Voltage Ranges for 120 V

	Service Voltage		Utilization Voltage	
	Minimum	Maximum	Minimum	Maximum
Range A	114 (–5%)	126 (+5%)	110 (–8.3%)	125 (+4.2%)
Range B	110 (–8.3%)	127 (+5.8%)	106 (–11.7%)	127 (+5.8%)

these limits. As the standard says, voltage excursions "should be infrequent."

- *Range B* — These requirements are more relaxed than Range A limits. According to the standard: "Although such conditions are a part of practical operations, they shall be limited in extent, frequency, and duration. When they occur, corrective measures shall be undertaken within a reasonable time to improve voltages to meet Range A requirements." Utilization equipment should give acceptable performance when operating within the Range B utilization limits, "insofar as practical" according to the standard.

These limits only apply to sustained voltage levels and not to momentary excursions, sags, switching surges, or short-duration interruptions.

Table 5.1 shows the most important limits, the limits on low-voltage systems. The table is given on a 120-V base; it applies at 120 V but also to any low-voltage system up to and including 600 V. The main target for utilities is the Range A service voltage, 114 to 126 V.

ANSI C84.1 defines three voltage classes: low voltage (1 kV or less), medium voltage (greater than 1 kV and less than 100 kV), and high voltage (greater than or equal to 100 kV). Within these classes, ANSI provides standard nominal system voltages along with the voltage ranges. A more detailed summary of the ANSI voltages is shown in Table 5.2 and Table 5.3.

For low-voltage classes, two nominal voltages are given — one for the electric system and a second, somewhat lower, nominal for the utilization equipment (for low-voltage motors and controls; other utilization equipment may have different nominal voltages). In addition, the standard gives common nameplate voltage ratings of equipment as well as information on what nominal system voltages the equipment is applicable to. As the standard points out, there are many inconsistencies between equipment voltage ratings and system nominal voltages.

For medium-voltage systems, ANSI C84.1 gives tighter limits for Ranges A and B. Range A is –2.5 to +5%, and Range B is –5 to +5.8%. However, most utilities do not follow these as limits for their primary distribution systems (utilities use the ANSI service voltage guidelines and set their primary voltage limits to meet the service voltage guidelines based on their practices). The three-wire voltages of 4,160, 6,900, and 13,800 V are mainly suited for industrial customers with large motors. Industrial facilities use motors on these systems with ratings of 4,000, 6,600, and 13,200 V, respectively.

TABLE 5.2

ANSI Standard Nominal System Voltages and Voltage Ranges for
Low-Voltage Systems

		Range A			Range B		
		Maximum	Minimum		Maximum	Minimum	
Nominal System Voltage	Nominal Utilization Voltage	Utilization and Service Voltage[a]	Service Voltage	Utilization Voltage	Utilization and Service Voltage	Service Voltage	Utilization Voltage
Two Wire, Single Phase							
120	115	126	114	110	127	110	106
Three Wire, Single Phase							
120/240	115/230	126/252	114/228	110/220	127/254	110/220	106/212
Four Wire, Three Phase							
208Y/120	200	218/126	197/114	191/110	220/127	191/110	184/106
240/120	230/115	252/126	228/114	220/110	254/127	220/110	212/106
480Y/277	460	504/291	456/263	440/254	508/293	440/254	424/245
Three Wire, Three Phase							
240	230	252	228	220	254	220	212
480	**460**	**504**	**456**	**440**	**508**	**440**	**424**
600	575	630	570	550	635	550	530

Note: Bold entries show preferred system voltages.

[a] The maximum utilization voltage for Range A is 125 V or the equivalent (+4.2%) for other nominal voltages through 600 V.

Improper voltage regulation can cause many problems for end users. Sustained overvoltages or undervoltages can cause the following end-use impacts:

- *Improper or less-efficient equipment operation* — For example, lights may give incorrect illumination or a machine may run fast or slow.
- *Tripping of sensitive loads* — For example, an uninterruptible power supply (UPS) may revert to battery storage during high or low voltage. This may drain the UPS batteries and cause an outage to critical equipment.

In addition, undervoltages can cause

- *Overheating of induction motors* — For lower voltage, an induction motor draws higher current. Operating at 90% of nominal, the full-load current is 10 to 50% higher, and the temperature rises by 10 to 15%. With less voltage, the motor has reduced motor starting torque.

TABLE 5.3

ANSI Standard Nominal System Voltages and Voltage Ranges for
Medium-Voltage Systems

	Range A			Range B		
	Maximum	Minimum		Maximum	Minimum	
Nominal System Voltage	Utilization and Service Voltage	Service Voltage	Utilization Voltage	Utilization and Service Voltage	Service Voltage	Utilization Voltage
Four Wire, Three Phase						
4160Y/2400	4370/2520	4050/2340	3740/2160	4400/2540	3950/2280	3600/2080
8320Y/4800	8730/5040	8110/4680		8800/5080	7900/4560	
12000Y/6930	12600/7270	11700/6760		12700/7330	11400/6580	
12470Y/7200	**13090/7560**	**12160/7020**		**13200/7620**	**11850/6840**	
13200Y/7620	13860/8000	12870/7430		13970/8070	12504/7240	
13800Y/7970	14490/8370	13460/7770		14520/8380	13110/7570	
20780Y/1200	21820/12600	20260/11700		22000/12700	19740/11400	
22860Y/13200	24000/13860	22290/12870		24200/13970	21720/12540	
24940Y/14400	**26190/15120**	**24320/14040**		**26400/15240**	**23690/13680**	
34500Y/19920	**36230/20920**	**33640/19420**		**36510/21080**	**32780/18930**	
Three Wire, Three Phase						
2400	2520	2340	2160	2540	2280	2080
4160	**4370**	**4050**	**3740**	**4400**	**3950**	**3600**
4800	5040	4680	4320	5080	4560	4160
6900	7240	6730	6210	7260	6560	5940
13800	**14490**	**13460**	**12420**	**14520**	**13110**	**11880**
23000	24150	22430		24340	21850	
34500	36230	33640		36510	32780	

Notes: Bold entries show preferred system voltages. Some utilization voltages are blank because utilization equipment normally does not operate directly at these voltages.

Also, overvoltages can cause

- *Equipment damage or failure* — Equipment can suffer insulation damage. Incandescent light bulbs wear out much faster at higher voltages.

- *Higher no-load losses in transformers* — Magnetizing currents are higher at higher voltages.

5.2 Voltage Drop

We can approximate the voltage drop along a circuit as

$$V_{drop} = |V_s| - |V_r| \approx I_R \cdot R + I_X \cdot X$$

where
  V_{drop} = voltage drop along the feeder, V
  R = line resistance, Ω
  X = line reactance, Ω
  I_R = line current due to real power flow (in phase with the voltage), A
  I_X = line current due to reactive power flow (90° out of phase with the voltage), A
In terms of the load power factor, *pf*, the real and reactive line currents are

$$I_R = I \cdot pf = I \cos \theta$$

$$I_X = I \cdot qf = I \sin \theta = I \sin(\cos^{-1}(pf))$$

where
  I = magnitude of the line current, A
  pf = load power factor
  qf = load reactive power factor = $\sin(\cos^{-1}(pf))$
  θ = angle between the voltage and the current

While just an approximation, Brice (1982) showed that $I_R \cdot R + I_X \cdot X$ is quite accurate for most distribution situations. The largest error occurs under heavy current and leading power factor. The approximation has an error less than 1% for an angle between the sending and receiving end voltages up to 8° (which is unlikely on a distribution circuit). Most distribution programs use the full complex phasor calculations, so the error is mainly a consideration for hand calculations.

This approximation highlights two important aspects about voltage drop:

- *Resistive load* — At high power factors, the voltage drop strongly depends on the resistance of the conductors. At a power factor of 0.95, the reactive power factor (*qf*) is 0.31; so even though the resistance is normally smaller than the reactance, the resistance plays a major role.

- *Reactive load* — At moderate to low power factors, the voltage drop depends mainly on the reactance of the conductors. At a power factor of 0.8, the reactive power factor is 0.6, and because the reactance is usually larger than the resistance, the reactive load causes most of the voltage drop. Poor power factor significantly increases voltage drop.

Voltage drop is higher with lower voltage distribution systems, poor power factor, single-phase circuits, and unbalanced circuits. The main ways to reduce voltage drop are to:

- Increase power factor (add capacitors)
- Reconductor with a larger size

- Balance circuits
- Convert single-phase sections to three-phase sections
- Reduce load
- Reduce length

In many cases, we can live with significant voltage drop as long as we have enough voltage regulation equipment to adjust for the voltage drop on the circuit.

5.3   Regulation Techniques

Distribution utilities have several ways to control steady-state voltage. The most popular regulation methods include:

- Substation load tap-changing transformers (LTCs)
- Substation feeder or bus voltage regulators
- Line voltage regulators
- Fixed and switched capacitors

Most utilities use LTCs to regulate the substation bus and supplementary feeder regulators and/or switched capacitor banks where needed.

Taps on distribution transformers are another tool to provide proper voltage to customers. Distribution transformers are available with and without no-load taps (meaning the taps are to be changed without load) with standard taps of ±2.5 and ±5%. Utilities can use this feature to provide a fixed boost for customers on a circuit with low primary voltage. This also allows the primary voltage to go lower than most utilities would normally allow. Remember, the service entrance voltage is most important. Most distribution transformers are sold without taps, so this practice is not widespread. It also requires consistency; an area of low primary voltage may have several transformers to adjust — if one is left out, the customers fed by that transformer could receive low voltage.

5.3.1   Voltage Drop Allocation and Primary Voltage Limits

Most utilities use the ANSI C84.1 ranges for the service entrance, 114 to 126 V. How they control voltage and allocate voltage drop varies. Consider the voltage profile along the circuit in Figure 5.1. The substation LTC or bus regulator controls the voltage at the source. Voltage drops along the primary line, the distribution transformer, and the secondary. We must consider the customers at the start and end of the circuit:

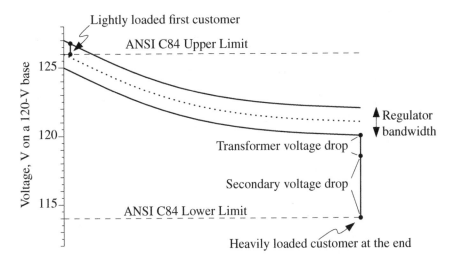

FIGURE 5.1
Voltage drop along a radial circuit with no capacitors or line regulators.

- *End — Heavily loaded* — Low voltages are a concern, so we consider a heavily loaded transformer and secondary. The allocation across the secondary depends on the utility's design practices as far as allowable secondary lengths and conductor sizes are concerned.

- *Source — Lightly loaded* — Near the source, we can operate the primary above 126 V, but we must ensure that the first customer does not have overvoltages when that customer is lightly loaded. Commonly, utilities assume that the secondary and transformer drop to this lightly loaded customer is 1 V. With that, the upper primary voltage limit is 127 V.

In the voltage drop along the primary, we must consider the regulator bandwidth (and bandwidths for capacitors if they are switched based on voltage). Voltage regulators allow the voltage to deviate by half the bandwidth in either direction. So, if we have a 2-V bandwidth and a desired range of 7 V of primary drop, subtracting the 2-V bandwidth only leaves 5 V of actual drop (see Figure 5.1). Likewise, if we choose 127 V as our upper limit on the primary, our maximum set voltage is 126 V with a 2-V regulator bandwidth.

Normally, utilities use standardized practices to allocate voltage drop. Deviations from the standard are possible but often not worth the effort.

If we have an express feeder at the start of a circuit, we can regulate the voltage much higher than 126 V as long as the voltage drops enough by the time the circuit reaches the first customer.

Primary voltage allocation affects secondary allocation and vice versa. A rural utility may have to allow a wide primary voltage range to run long

TABLE 5.4

Primary Voltage Ranges at Several Utilities

Service Area Type	Minimum	Maximum	Percent Range
Dense urban area	120	127	5.4
Dense urban area	117	126	7.5
Urban/suburban	114	126	10.0
Urban/suburban	115	125	8.3
Urban/suburban			
No conservation reduction	119	126	5.8
With conservation reduction	119	123	3.3
Multi-state area	117	126	7.5
Multi-state area			
Urban standard	123	127	3.3
Rural standard	119	127	6.6
Suburban and rural	113	125	10.0
Suburban and rural			
Urban standard	116	125	7.5
Rural standard	112	125	10.8
Urban and rural	115	127	10.0
Rural, mountainous	116	126	8.3
Rural, mountainous	113	127	11.7

Source: Willis, H. L., *Power Distribution Planning Reference Book*, Marcel Dekker, New York, 1997b, with additional utilities added.

circuits, which leaves little voltage drop left for the transformer and secondary. Since rural loads are typically each fed by their own transformer, rural utilities can run the primary almost right to the service entrance. Using low-impedance distribution transformers and larger-than-usual transformers also helps reduce the voltage drop beyond the primary. For the secondary conductors, triplex instead of open wire and larger size conductors help reduce secondary drop. Utilities that allow less primary voltage drop can run longer secondaries.

Utility practices on voltage limits on the primary range widely, as shown in Table 5.4. The upper range is more consistent — most are from 125 to 127 V — unless the utility uses voltage reduction (for energy conservation or peak shaving). The lower range is more variable, anywhere from 112 to 123 V. Obviously, the utility that uses a 112-V lower limit is not required to abide by the ANSI C84.1 limits.

5.3.2 Load Flow Models

Load flows provide voltage profiles that help when planning new distribution circuits, adding customers, and tracking down and fixing voltage problems. Most distribution load-flow programs offer a function to plot the voltage as a function of distance from the source.

We can model a distribution circuit at many levels of detail. Many utilities are modeling more of their systems in more detail. For most load flows, utilities normally just model the primary. Modeling the secondary is occasionally useful for modeling specific problems at a customer. We can still have very good models with simplifications. Modeling long laterals or branches is normally a good idea, but we can lump most laterals together as a load where they tie into the main line. Modeling each transformer as a load is rarely worth the effort; we can combine loads together and maintain accuracy with some common sense. Most mainline circuits can be accurately modeled if broken into 10 to 20 sections with load lumped with each section. Of course, accurate models of capacitors and line regulators are a good idea.

Correctly modeling load phasing provides a better voltage profile on each phase. Unbalanced loads cause more voltage drop because of:

- *Higher loop impedance* — The impedance seen by unbalanced loads, the loop impedance including the zero-sequence impedance, is higher than the positive-sequence impedance seen by balanced loads.

- *Higher current on the loaded phases* — If the current splits unevenly by phases, the more heavily loaded phases see more voltage drop.

Utilities often do not keep accurate phasing information, but it helps improve load-flow results. We do not need the phasing on every transformer, but we will have better accuracy if we know the phasing of large single-phase taps.

Of the data entered into the load flow model, the load allocation is the trickiest. Most commonly, loads are entered in proportion to the transformer kVA. If a circuit has a peak load equal to the sum of the kVA of all of the connected transformers divided by 2.5, then each load is modeled as the given transformer size in kVA divided by 2.5. Incorporating metering data is another more sophisticated way to allocate load. If a utility has a transformer load management system or other system that ties metered kilowatt-hour usage to a transformer to estimate loadings, feeding this data to the load flow can yield a more precise result. In most cases, all of the loads are given the same power factor, usually what is measured at the substation. Additional measurements could be used to fine-tune the allocation of power factor. Some utilities also assign power factor by customer class.

Most distribution load flow programs offer several load types, normally constant power, constant current, and constant impedance:

- *Constant power load* — The real and reactive power stays constant as the voltage changes. As voltage decreases, this load draws more current, which increases the voltage drop. A constant power model is good for induction motors.

- *Constant current load* — The current stays constant as the voltage changes, and the power increases with voltage. As voltage decreases, the current draw stays the same, so the voltage drop does not change.

TABLE 5.5

Load Modeling Approximations Recommended by Willis (1997a)

Feeder Type	Percent Constant Power	Percent Constant Impedance
Residential and commercial, summer peaking	67	33
Residential and commercial, winter peaking	40	60
Urban	50	50
Industrial	100	0
Developing countries	25	75

Source: Willis, H. L., "Characteristics of Distribution Loads," in *Electrical Transmission and Distribution Reference Book.* Raleigh, NC, ABB Power T&D Company, 1997.

- *Constant impedance load* — The impedance is constant as the voltage changes, and the power increases as the square of the voltage. As voltage decreases, the current draw drops off linearly; so the voltage drop decreases. The constant impedance model is good for incandescent lights and other resistive loads.

Normally, we can model most circuits as something like 40 to 60% constant power and 40 to 60% constant impedance (see Table 5.5 for one set of recommendations). Modeling all loads as constant current is a good approximation for many circuits. Modeling all loads as constant power is conservative for voltage drop.

5.3.3   Voltage Problems

Voltage complaints (normally undervoltages) are regular trouble calls for utilities. Some are easy to fix; others are not. First, check the secondary. Before tackling the primary, confirm that the voltage problem is not isolated to the customers on the secondary. If secondary voltage drop is occurring, check loadings, make sure the transformer is not overloaded, and check for a loose secondary neutral.

If the problem is on the primary, some things to look for include:

- *Excessive unbalance* — Balancing currents helps reduce voltage drop.
- *Capacitors* — Look for blown fuses, incorrect time clock settings, other incorrect control settings, or switch malfunctions.
- *Regulators* — Check settings. See if more aggressive settings can improve the voltage profile enough: a higher set voltage, more line drop compensation, and/or a tighter bandwidth.

These problems are relatively easy to fix. If it is not these, and if there is too much load for the given amount of impedance, we will have to add equipment to fix the problem. Measure the primary voltage (and if possible the loadings) at several points along the circuit. An easy way to measure the primary voltage is to find a lightly loaded distribution transformer and measure the secondary voltage. Measure the power factor at the substation. A poor power factor greatly increases the voltage drop.

Load flows are a good tool to try out different options to improve voltage on a circuit. If possible, match voltage profiles with measurements on the circuit. Measurements provide a good sanity check. Try to measure during peak load conditions. Regulator and capacitor controllers can provide extra information if they have data logging capability. Normally, we allocate the load for the model equally by transformer kVA. This may not always be right, and measurements can help "tweak" the model. A load flow can help determine the best course of action. Where do we need a supplementary line regulator? How many? Can fixed capacitors do the job? Do we need switched capacitors? Circuits with poor power factor are the best candidates for capacitors as they will help reduce line losses as well as improve voltage.

In addition to extra regulating equipment, consider other options. Sometimes, we can move one or more circuit sections to a different feeder to reduce the loading on the circuit. If transformers have taps, investigate changing the transformer taps. Though it is expensive, we can also build new circuits, upgrade to a higher voltage, or reconductor.

5.3.4 Voltage Reduction

Utilities can use voltage adjustments as a way to manage system load. Voltage reduction can reduce energy consumption and/or reduce peak demand. Several studies have shown roughly a linear response relationship between voltage and energy use — a 1% reduction in voltage reduces energy usage by 1% (or just under 1%, depending on the study). Kirshner and Giorsetto (1984) analyzed trials of conservation voltage reduction (CVR) at several utilities. While results varied significantly, most test circuits had energy savings of between 0.5 and 1% for each 1% voltage reduction. Their regression analysis of the feeders found that residential energy savings were 0.76% for each 1% reduction in voltage, while commercial and industrial loads had reductions of 0.99% and 0.41% (but the correlations between load class and energy reduction were fairly small).

Voltage reduction works best with resistive loads because the power drawn by a resistive load decreases with the voltage squared. Lighting and resistive heating loads are the dominant resistive loads; these are not ideal resistive loads. For example, the power on incandescent lights varies as the voltage to the power of about 1.6, which is not quite to the power of 2 but close. Residential and commercial loads have higher percentages of resistive load. For water heaters and other devices that regulate to a temperature, reducing voltage does not reduce overall energy usage; the devices just run more often.

Voltage reduction to reduce demand has even more impact than that on energy reduction. The most reduction occurs right when the voltage is reduced, and then some of the reduction is lost as some loads keep running longer than normal to compensate for lower voltage. For example, Priess and Warnock (1978) found that during a 4-h, 5% voltage reduction, the demand on one typical residential circuit dropped by 4% initially and diminished to a 3% drop by the end of the 4-h period.

Voltage reduction works best on short feeders — those that do not have much voltage drop. On these, we can control reduction just through adjustments of the station LTC regulator settings. It is straightforward to set up a system where operators can change the station set voltage through SCADA. On longer circuits, we need extra measures. Some strategies include:

- *Extra regulators* — Extra regulators can help flatten the voltage profile along the circuit. Each regulator is set with a set voltage and compensation settings appropriate for a tighter voltage range. This approach is most appropriate for energy conservation. Controlling the regulators to provide peak shaving is difficult; the communications and controls add significantly to the cost.

- *Feeder capacitors* — The vars injected by capacitors help flatten the voltage profile and allow a lower set voltage on the station LTC. On many circuits, just fixed capacitors can flatten the profile enough to reduce the station set voltage. McCarthy (2000) reported how Georgia Power used this strategy to reduce peak loads by 500 kW on circuits averaging approximately 18 MW.

- *Tighter bandwidth* — With a smaller regulator bandwidth, the voltage spread on the circuit is smaller. A smaller bandwidth requires more frequent regulator or LTC maintenance (the regulator changes taps more often) but not drastic differences. Kirshner (1990) reported that reducing the bandwidth from 3 to 1.5 V doubled the number of regulator tap changes.

- *Aggressive line drop compensation* — An aggressive line-drop compensation scheme can try to keep the voltage at the low end (say, at 114 V) for the last customer at all times. The set voltage in the station may be 115 to 117 V, depending on the circuit voltage profile. Aggressive compensation boosts the voltage during heavy loads, while trying to keep voltages low at the ends of circuits. During light loads, the station voltage may drop to well under 120 V. This strategy helps the least at heavy load periods, so it is more useful for energy conservation than for peak shaving. Aggressive compensation makes low voltages more likely at the end of circuits. If any of the planning assumptions are wrong, especially power factor and load placement, customers at the end of circuits can have low voltages.

- *Others* — Other voltage profile improvement options help when implementing a voltage reduction program, although some of these options, such as reconductoring, undergrounding, load balancing, and increasing primary voltage levels, are quite expensive.

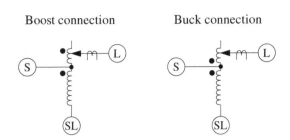

FIGURE 5.2
ANSI type A single-phase regulator, meaning taps on the load bushing.

5.4 Regulators

Voltage regulators are autotransformers with automatically adjusting taps. Commonly, regulators provide a range from −10 to +10% with 32 steps. Each step is 5/8%, which is 0.75 V on a 120-V scale.

A single-phase regulator has three bushings: the source (S), the load (L), and the source-load (SL). The series winding is between S and L. Figure 5.2 shows a straight regulator (ANSI type A) with the taps on the load side. An ANSI type B, the inverted design, has the taps on the source bushing. The regulator controller measures current with a CT at the L bushing and measures the voltage with a PT between L and SL. Regulators have a reversing switch that can flip the series winding around to change back and forth between the boost and the buck connection.

Regulators are rated on current (IEEE Std. C57.15-1999). Regulators also have a kVA rating which is the two-winding transformer rating and not the load-carrying capability. A regulator at 7.62 kV line to ground with a ±10% range and a load current rating of 100 A has a kVA rating of 0.1(7.62 kV)(100A) = 76 kVA. The load-carrying capability is ten times the regulator's kVA rating.

By reducing the range of regulation, we can extend the rating of the regulator. Reducing the range from ±10 to ±5% increases the rating by 60% (see Figure 5.3).

The impedance is the two-winding impedance times a base value about ten times as large. Because the impedance is so small, we can normally neglect it.

Three-phase regulators, often used in stations, are used on wye or delta systems. A three-phase regulator controls all three phases simultaneously. These are normally larger units. The normal connection internally is a wye connection with the neutral point floating.

Commonly, utilities use single-phase units, even for regulating three-phase circuits. We can connect single-phase regulators in several ways [see Figure 5.4 and (Bishop et al., 1996)]:

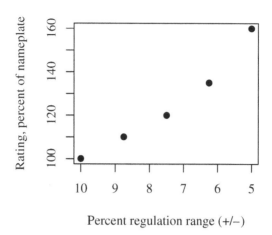

FIGURE 5.3
Increased regulator ratings with reduced regulation range.

Grounded-wye connection

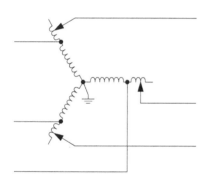

Open-delta connection   Closed-delta (leading) connection

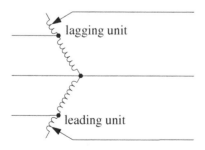

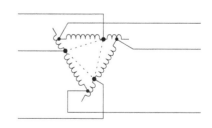

FIGURE 5.4
Three-phase regulator connections.

- *Line to neutral* — On four-wire systems, three-phase circuits normally have three single-phase regulators connected line to neutral. Line-to-neutral connections are also appropriate for single-phase and two-phase circuits. Each regulator independently controls voltage, which helps control voltage unbalance as well as steady-state voltage.

- *Open delta* — Only two single-phase regulators are needed, each connected phase to phase.

- *Closed delta* — Three regulators are connected phase to phase. Using the closed delta extends the regulation range by 50%, from ±10 to ±15%.

In both of the delta connections, the regulators see a current phase-shifted relative to the voltage. In the leading connection with unity power factor loads, the line current through the regulator leads the line-to-line voltage by 30°. The lagging connection has the current reversed: for a unit power factor load, the line current lags the line-to-line voltage by 30°. In the open-delta configuration, one of the units is leading and the other is lagging. In the closed-delta arrangement, all three units are either leading or all three are lagging. Although uncommon, both of the delta connections can be applied on four-wire systems.

Regulators have a voltage regulating relay that controls tap changes. This relay has three basic settings that control tap changes (see Figure 5.5):

- *Set voltage* — Also called the set point or bandcenter, the set voltage is the desired output of the regulator.

- *Bandwidth* — Voltage regulator controls monitor the difference between the measured voltage and the set voltage. Only when the difference exceeds one half of the bandwidth will a tap change start. Use a bandwidth of at least two times the step size, 1.5 V for ±10%, 32-step regulators. Settings of 2 and 2.5 V are common.

- *Time delay* — This is the waiting time between the time when the voltage goes out of band and when the controller initiates a tap

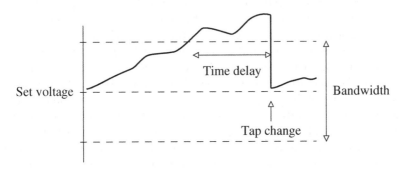

FIGURE 5.5
Regulator tap controls based on the set voltage, bandwidth, and time delay.

change. Longer time delays reduce the number of tap changes. Typical time delays are 30 to 60 sec.

If the voltage is still out of bounds after a tap change, the controller makes additional tap changes until the voltage is brought within bounds. The exact details vary by controller, and some provide programmable modes. In some modes, controllers make one tap change at a time. In other modes, the controller may initiate the number of tap changes it estimates are needed to bring the voltage back within bounds. The time delay relay resets if the voltage is within bounds for a certain amount of time.

A larger bandwidth reduces the number of tap changes, but at a cost. With larger bandwidth, the circuit is not as tightly regulated. We should include the bandwidth in voltage profile calculations to ensure that customers are not given over or under voltages. Voltage that was used for bandwidth can be used for voltage drop along the circuit. With a higher bandwidth we may need more regulators on a given line. So, use at least two times the step size, but do not use excessively high bandwidths such as 3 or 3.5 V.

In addition to these basics, regulator controllers also have line-drop compensation to boost voltages more during heavy load. Controllers also may have high and low voltage limits to prevent regulation outside of a desired range of voltages. In addition to the regulator and control application information provided here, see Beckwith (1998), Cooper Power Systems (1978), General Electric (1979), and Westinghouse (1965).

Many regulators are bi-directional units; they can regulate in either direction, depending on the direction of power flow. A bi-directional regulator measures voltage on the source side using an extra PT or derives an estimate from the current. If the regulator senses reverse power flow, it switches to regulating the side that is normally the source side. We need reverse mode for a regulator on circuits that could be fed by an alternate source in the reverse direction. Without a reverse mode, the regulator can cause voltage problems during backfeeds. If a unidirectional regulator is fed "backwards," the regulator PT is now on the side of the source. Now, if the voltage drops, the regulator initiates a tap raise. However, the voltage the PT sees does not change because it is on the source side (very stiff). What happened was the voltage on the load side went down (but the regulator controller does not know that because it is not measuring that side). The controller still sees low voltage, so it initiates another tap raise which again lowers the voltage on the other side of the regulator. The controller keeps trying to raise the voltage until it reaches the end of its regulation range. So, we have an already low voltage that got dropped by an extra 10% by the unidirectional regulator. If the controller initially sees a voltage above its set voltage, it ratchets all the way to the high end causing a 10% overvoltage. Also, if the incoming voltage varies above and below the bandwidth, the regulator can run back and forth between extremes. A bi-directional regulator prevents these runaways. Depending on its mode, under reverse power, a bi-directional regulator can regulate in the reverse direction, halt tap changes, or move to the neutral point (these last two do not require PTs on both sides but just power direction sensing).

Regulators also have an operations counter. The counter helps identify when a regulator is due for refurbishment. Regulators are designed to perform many tap changes, often over one million tap changes over the life of a regulator. A regulator might change taps 70 times per day, which is 25,000 times per year (Sen and Larson, 1994). A regulator counter also provides a good warning indicator; excessive operations suggest that something is wrong, such as wrong line drop compensation settings, a bandwidth or time delay that is too small, or widely fluctuating primary voltages.

Regulators have "drag hands" — markers on the tap position indicator that show the maximum and minimum tap positions since the drag hands were last reset. The drag hands are good indicators of voltage problems. If maintenance reviews continually show the drag upper hand pegging out at +10%, the upstream voltage is probably too low. More work is needed to correct the circuit's voltage profile. Advanced controllers record much more information, including tap change records and demand metering to profile voltages, currents, and power factors.

5.4.1 Line-Drop Compensation

LTC transformer and regulator controls can be augmented with line-drop compensation. During heavy load, the controller boosts voltage the most, and during light load, voltage is boosted the least. The line-drop compensator uses an internal model of the impedance of the distribution line to match the line impedance. The user can set the R and X values in the compensator to adjust the compensation. The controller adjusts taps based on the voltage at the voltage regulating relay, which is the PT voltage plus the voltage across the line-drop compensator circuit (see Figure 5.6). With no compensation, the voltage regulating relay adjusts the taps based on the PT voltage.

Since load on a typical distribution line is distributed, R and X compensator settings are chosen so that the maximum desired boost is obtained

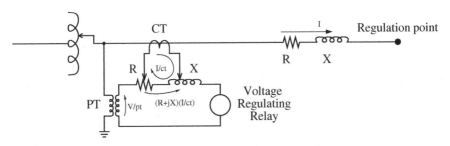

FIGURE 5.6
Line drop compensator circuit.

under heavy load while a given voltage is obtained under light load. There are two main approaches for selecting settings:

- *Load center* — The settings are chosen to regulate the voltage at a given point downstream of the regulator.
- *Voltage spread* — The R and X settings are chosen to keep the voltage within a chosen band when operating from light load to full load. The R and X settings may or may not be proportional to the line's R and X.

The main complication of all of the methods is that the load and power factors change (especially with downstream capacitor banks). Many regulators are set up without line drop compensation. It is obviously easier and less prone to mistakes, but we are losing out on some significant capability. If we set the regulator set voltage at 120 V, and we do not get enough boost along the line, we will need more regulators. With a higher set voltage such as 126 V, we do not need as many regulators, but we have high voltages at light load and possibly overvoltages if the circuit has capacitors. With line drop compensation, we have boost when we need it during heavy load, but not during light load (see Figure 5.7). Line-drop compensation also normally leads to a smaller range of fluctuations in voltage through the day for customers along the circuit.

5.4.1.1 Load-Center Compensation

The classic way to set compensator settings is to use the *load-center* method. Consider a line with impedances R_L and X_L with a load at the end. Now, if we pick the R_{set} and X_{set} of the compensator to match those of the line, as the load changes the regulator responds and adjusts the regulator taps to keep the voltage constant, not at the regulator but at the load. To achieve this, we can set the R_{set} and X_{set} of the regulator as

$$R_{set} = \frac{I_{CT}}{N_{PT}} R_L$$

$$X_{set} = \frac{I_{CT}}{N_{PT}} X_L$$

where
  R_{set} = regulator setting for resistive compensation, V
  X_{set} = regulator setting for reactive compensation, V
  I_{CT} = primary rating of the current transformer, A
  N_{PT} = potential transformer ratio (primary voltage/secondary voltage)
  R_L = primary line resistance from the regulator to the regulation point, Ω
  X_L = primary line reactance from the regulator to the regulation point, Ω

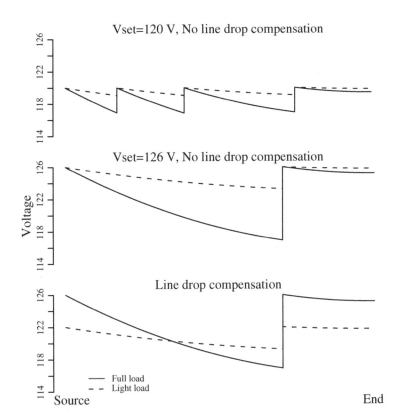

FIGURE 5.7
Voltage profiles on a circuit with various forms of regulation.

A regulator's R and X compensator settings are in units of volts. By using volts as units, we can directly see the impact of the regulator on a 120-V scale. Consider an example where the set voltage is 120 V. With a current at unity power factor and $R_{set} = 6$ V (X_{set} does not matter at unity power factor), the controller regulates the voltage to $120 + 6 = 126$ V when the current is at the peak CT rating. If the current is at half of the CT rating, the controller regulates to the set voltage plus 3 or 123 V. Available compensator settings are normally from −24 to +24 V.

Note that the primary CT rating is an important part of the conversion to compensator settings. The CT rating may be the same as the regulator rating or it may be higher. The CT rating is given on the nameplate. Table 5.6 shows the regulator ratings and primary CT current rating for one manufacturer. Regulators may be applied where the nameplate voltage does not match the system voltage if they are close enough to still allow the desired regulation range at the given location. Also, some regulators have taps that allow them to be used at several voltages. Make sure to use the appropriate PT ratio for the tap setting selected.

TABLE 5.6

Regulator and Primary CT Ratings in Amperes

Regulator Current Ratings	CT Primary Current
25	25
50	50
75	75
100	100
150	150
167, 200	200
219, 231, 250	250
289, 300	300
328, 334, 347, 400	400
418, 438, 463, 500	500
548, 578, 656, 668	600
833, 875, 1000, 1093	1000
1332, 1665	1600

When specifying impedances for the line-drop compensator, use the correct line impedances. For a three-phase circuit, use the positive-sequence impedance. For a single-phase line, use the loop impedance Z_S which is about twice the positive-sequence impedance.

On a delta regulator, either an open delta or a closed delta, divide the PT ratio by $\sqrt{3}$. On a delta regulator the PT connects from phase to phase, but the internal circuit model of the line-drop compensator is phase to ground, so we need the $\sqrt{3}$ factor to correct the voltage.

Line-drop compensation works perfectly for one load at the end of a line, but how do we set it for loads distributed along a line? If loads are uniformly distributed along a circuit that has uniform impedance, we can hold the voltage constant at the midpoint of the section by using:

- *3/8 rule* — For a uniformly distributed load, a regulator can hold the voltage constant at the midpoint of the circuit if we use line-drop compensation settings based on 3/8 of the total line impedance. A circuit with a uniformly distributed load has a voltage drop to the end of the circuit of one half of the drop had all of the loads been lumped into one load at the end of the circuit. Three-fourths of this drop is on the first half of the circuit, so (1/2)(3/4) = 3/8 is the equivalent voltage drop on a uniformly distributed load.

Make sure not to allow excessive voltages. We can only safely compensate a certain amount, and we will have overvoltages just downstream of the regulator if we compensate too much. Check the voltage to the voltage regulating relay to ensure that it is not over limits. The maximum voltage is

$$V_{max} = V_{set} + (pf \cdot R_{set} + qf \cdot X_{set}) \, I_{max}$$

where

V_{set} = regulator set voltage
R_{set} = resistive setting for compensation, V
X_{set} = reactive setting for compensation, V
pf = load power factor
qf = load reactive power factor = $\sin(\cos^{-1}(pf))$
I_{max} = maximum load current in per unit relative to the regulator CT rating

If V is more than what you desired, reduce R_{set} and X_{set} appropriately to meet your desired limit.

5.4.1.2   Voltage-Spread Compensation

In another method, the *voltage-spread* method, we find compensator settings by specifying the band over which the load-side voltage should operate. For example, we might want the regulator to regulate to 122 V at light load and 126 V at full load. If we know or can estimate the light-load and full-load current, we can find R and X compensator settings to keep the regulated voltage within the proper range. If we want the regulator to operate over a given compensation range C, we can choose settings to satisfy the following:

$$C = V - V_{set} = pf \cdot R_{set} + qf \cdot X_{set}$$

where

R_{set} = resistive setting for compensation, V
X_{set} = reactive setting for compensation, V
pf = load power factor
qf = load reactive power factor = $\sin(\cos^{-1}(pf))$
C = total desired compensation voltage, V
V_{set} = regulator set voltage, V
V = voltage that the controller will try to adjust the regulator to, V

With line current operating to the regulator CT rating limit (which is often the regulator size) and the current at the given power factor, these settings will boost the regulator by C volts on a 120-V scale. Any number of settings for R_{set} and X_{set} are possible to satisfy this equation. If we take $X_{set} = \frac{X}{R} R_{set}$ where the X/R ratio is selectable, the settings are

$$R_{set} = \frac{C}{pf + \frac{X}{R} qf}$$

$$X_{set} = \frac{\frac{X}{R} C}{pf + \frac{X}{R} qf} = \frac{X}{R} R_{set}$$

where

$\frac{X}{R}$ = X/R ratio of the compensator settings

Note that C must be given as seen on the regulator PT secondaries, on a 120-V base. As an example, if the feeder voltage should be not more than 126 V at the limit of the regulator, and the desired voltage at no load is 122 V, set the regulator set voltage at 122 V and find R_{set} and X_{set} to give $C = 4$ V. For a power factor of 0.85 and $\frac{X}{R} = 3$, the equations above give $R_{set} = 1.64$ V and $X_{set} = 4.94$ V.

To control the voltage range for a light load other than zero and for a peak load other than the regulator CT rating, we can use the following to find the voltage swing from light load to full load as

$$V_{max} - V_{min} = (pf \cdot R_{set} + qf \cdot X_{set})I_{max} - (pf \cdot R_{set} + qf \cdot X_{set})I_{min}$$

where

V_{max} = desired voltage at the maximum load current on a 120-V base, V
V_{min} = desired voltage at the minimum load current on a 120-V base, V
I_{max} = maximum load current in per-unit relative to the regulator CT rating
I_{min} = minimum load current in per-unit relative to the regulator CT rating

Now, the R and X settings are

$$R_{set} = \frac{V_{max} - V_{min}}{(pf + \frac{X}{R}qf)(I_{max} - I_{min})}$$

$$X_{set} = \frac{X}{R} R_{set}$$

And, the regulator set voltage is

$$V_{set} = V_{min} - (pf \cdot R_{set} + qf \cdot X_{set})I_{min} = V_{min} - \frac{V_{max} - V_{min}}{I_{max} - I_{min}} I_{min}$$

With a compensator X/R ratio equal to the line X/R ratio, these equations move the effective load center based on the choice of voltage and current minimums and maximums.

Just like we can choose to have the compensator X/R ratio equal the line X/R ratio, we can choose other values as well. There are good reasons why we might want to use other ratios; this is done mainly to reduce the sensitivity to power factor changes. The *zero reactance* method of selecting compensator makes $X_{set} = 0$ (and the compensator $X/R = 0$) but otherwise uses the same equations as the voltage spread method (General Electric, 1979).

By making X_{set} zero, the compensator is not sensitive to variations in power factor caused by switched capacitors or load variation; only real power changes cause regulator movement. This method also simplifies application of regulators. The equations become

$$R_{set} = \frac{V_{max} - V_{min}}{pf(I_{max} - I_{min})}$$

$$X_{set} = 0$$

And, the regulator set voltage is

$$V_{set} = V_{min} - (pf \cdot R_{set})I_{min}$$

The equations simplify more if we assume that $I_{min} = 0$ (our error with this is that voltages run on the high side during light load). A further simplification is to assume that the power factor is one. If the power factor is less than that at full load, the regulator will not boost the voltage quite as much. Often, we do not know the power factor at the regulator location anyway.

This method is useful with switched capacitor banks close to the regulator. It does not perform well for low power factors if we have assumed a power factor near unity. With this control, the regulator will not provide enough boost with poor power-factor load.

Another option is to take $X/R = 0.6$, which weights the real power flow more than the reactive power flow, but not as extremely as the zero reactance compensation method. So, although the controller is somewhat desensitized to changes in power factor, the regulator provides some action based on reactive power. Figure 5.8 shows several X/R compensator settings chosen to provide an operating band from 121 V at light load to 127 V at full load. The settings were chosen based on a power factor of 0.9, and the curves show the voltage as the power factor varies. The middle graph with $X/R = 0.6$ performs well over a wide range of power factors. The graph on the left, where $X/R = 3$ which is the line X/R ratio, has the most variation with changes in power factor. If power factor is lower than we expected, the compensator will cause high voltages.

With $X/R = 0.6$ and $pf = 0.9$, the voltage spread equations are

$$R_{set} = 0.86 \frac{V_{max} - V_{min}}{(I_{max} - I_{min})}$$

$$X_{set} = 0.6R_{set}$$

And, the regulator set voltage is

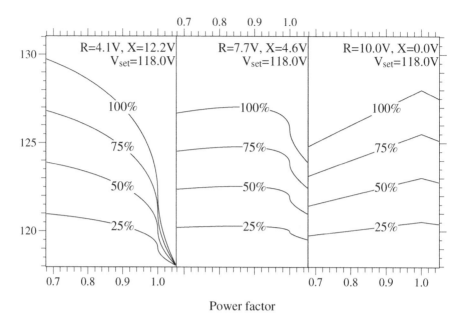

FIGURE 5.8
Regulated voltage based on different compensator settings and power factors with the percentage loadings given on the graph. All settings are chosen to operate from 121 V at light load (33%) to 127 V at full load (100% of the primary CT ratio) at a power factor of 0.9.

$$V_{set} = V_{min} - \frac{V_{max} - V_{min}}{I_{max} - I_{min}} I_{min}$$

The *universal compensator* method fixes compensation at R_{set} = 5 V and X_{set} = 3 V to give a 6-V compensation range with current ranging up to the regulator CT rating (General Electric, 1979). For other voltage ranges and maximum currents, we can use:

$$R_{set} = \frac{5}{I_{max}} \frac{(V_{max} - V_{min})}{6}$$

$$X_{set} = \frac{3}{I_{max}} \frac{(V_{max} - V_{min})}{6}$$

And we assume that I_{min} = 0, so the regulator set voltage is

$$V_{set} = V_{min}$$

To make this even more "cookbook," we can standardize on values of V_{max} and V_{min}, for example, values of 126 V and 120 V. If the full-load is the CT rating (which we might want in order to be conservative), the default settings become R_{set} = 5 V and X_{set} = 3 V. The universal compensation method is easy yet relatively robust.

With any of the voltage-spread methods of setting the R and X line-drop compensation, the peak current is an important parameter. If we underestimate the load current, the regulator can overcompensate and cause high voltages (if we do not have a voltage override limiter or if it is disabled). Check regulator loadings regularly to ensure that the compensation is appropriate.

5.4.1.3   Effects of Regulator Connections

On an open-delta regulator, one regulator is connected leading, and the other lagging. We need to adjust the compensator settings to account for the 30° phase shift. On the leading regulator, the current leads the voltage by 30°; so we need to subtract 30° from the compensator settings, which is the same as multiplying by $1\angle 30°$ or (cos 30° – j sin 30°). Modify the settings for the leading regulator (Cooper Power Systems, 1978; Westinghouse Electric Corporation, 1965) with

$$R'_{set} = 0.866\ R_{set} + 0.5X_{set}$$

$$X'_{set} = 0.866\ X_{set} - 0.5R_{set}$$

And for the lagging regulator we need to add 30°, which gives

$$R'_{set} = 0.866\ R_{set} - 0.5X_{set}$$

$$X'_{set} = 0.866\ X_{set} + 0.5R_{set}$$

For an X/R ratio above 1.67, R'_{set} is negative on the lagging regulator; and for a ratio below 0.58, X'_{set} is negative on the leading regulator. Most controllers allow negative compensation.

In the field, how do we tell between the leading and the lagging regulator? Newer regulator controllers can tell us which is which from phase angle measurements. For older controllers, we can modify the compensator settings to find out (Lokay and Custard, 1954). Set the resistance value on both regulators to zero, and set the reactance setting on both to the same nonzero value. The unit that moves up the most number of tap positions is the lagging unit (with balanced voltages, this is the unit that goes to the highest raise position). If the initial reactance setting is not enough, raise the reactance settings until the leading and lagging units respond differently.

With a closed-delta regulator, all three regulators are connected either leading or lagging. All three regulators have the same set of compensator settings; adjust them all with either the leading or the lagging equations described for the open-delta regulator.

On a three-phase regulator, even on a delta system, the compensator settings do not need adjustment. The controller accounts for any phase shift that might occur inside the regulator.

5.4.2 Voltage Override

Use the *voltage override* feature on the regulator controller. No matter how we select the line-drop compensation settings, an important feature is an upper voltage limit on the regulation action. The regulator keeps the regulated voltage below this limit regardless of the line-drop compensation settings. Always use this feature to protect against overvoltages caused by incorrect line-drop compensation settings or unusually high loadings. This upper voltage limiter is also called "first house protection," as it is the first few customers downstream that could have overvoltages due to regulator action. With a voltage limit, we can set line-drop compensator settings more aggressively and not worry about causing overvoltages to customers. On a regulator without an upper limit (normally older units), increase estimated peak loadings when calculating line-drop compensation settings in order to reduce the risk of creating overvoltages. Voltage override functions usually have a deadband type setting on the voltage limit to prevent repeated tap changes. For example, we might set a 126-V upper limit with a deadband of an extra 2 V. Above 128 V the controller immediately taps the regulator down to 126 V, and between 126 and 128 V the controller prohibits tap raises (different controllers implement this function somewhat differently; some include time delays). Even without line-drop compensation, the voltage override function helps protect against sudden changes in upstream voltages (the out-of-limit response is normally faster than normal time-delay settings programmed into regulators).

5.4.3 Regulator Placement

With no feeder regulators, the entire voltage drop on a circuit must be within the allowed primary voltage range. One feeder regulator can cover primary voltage drops up to twice the allowed voltage variation. Similarly, two supplementary regulators can cover primary voltage drops up to three times the allowed variation. For a uniformly distributed load, optimum locations for two regulators are at distances from the station of approximately 20% of the feeder length for one and 50% for the other. For one feeder regulator, the optimum location for a uniformly distributed load is at 3/8 of the line length from the station.

When placing regulators and choosing compensator settings, allow for some load growth on the circuit. If a regulator is applied where the load is right near its rating, it may not be able to withstand the load growth. However, it is more than just concern about the regulator's capability. If we want to keep the primary voltage above 118 V, and we add a regulator to a circuit right at the point where the primary voltage falls to 118 V, that will correct the voltage profile along the circuit with present loadings. If loadings increase in the future, the voltage upstream of the regulator will drop below 118 V. As previously discussed, when setting line-drop compensator settings, the maximum load on the regulator should allow room for load growth to reduce the chance that the regulator boosts the voltage too much.

Several regulators can be strung together on a circuit. Though this can meet the steady-state voltage requirements of customers, it will create a very weak source for them. Flicker problems from motors and other fluctuating loads are more likely.

Also consider the effect of dropped load on regulators. A common case is a recloser downstream of a line regulator. If the regulator is tapped up because of heavy load and the recloser suddenly drops a significant portion of the load, the voltage downstream of the regulator will pop up until the regulator controller shifts the taps back down.

5.4.4 Other Regulator Issues

Normally, voltage regulators help with voltage unbalance as each regulator independently controls its phase. If we aggressively compensate, the line-drop compensation can cause voltage unbalance. Consider a regulator set to operate between 120 V at no load and 126 V at full load. If one phase is at 50% load and the other two are at 0% load, the line-drop compensator will tap to 123 V on the loaded phase and to 120 V on the unloaded phases. Depending on customer placements, this may be fine if the voltages correct themselves along the line. But if the unbalance is due to a large tapped lateral just downstream of the regulator, the regulator needlessly unbalances the voltages.

Capacitor banks pose special coordination issues with regulators. A fixed capacitor bank creates a constant voltage rise on the circuit and a constant reactive contribution to the current. Either fixed or switched, capacitors upstream of a regulator do not interfere with the regulator's control action. Downstream capacitors pose the problem. A capacitor just downstream of a regulator affects the current that the regulator sees, but it does not measurably change the shape of the voltage profile beyond the regulator. In this case, we would like the line-drop compensation to ignore the capacitor. The voltage-spread compensation with a low compensator X/R or the zero-reactance compensator settings work well because they ignore or almost ignore the reactive current, so it works with fixed or switched banks downstream of the regulator. The load-center approach is more difficult to get to work with capacitors.

We do not want to ignore the capacitor at the end of a circuit section we are regulating because the capacitor significantly alters the profile along the circuit. In this case, we do not want zero-reactance compensation; we want some X to compensate for the capacitive current.

Switched capacitors can interact with the tap-changing controls on regulators upstream of the capacitors. This sort of interaction is rare but can happen if the capacitor is controlled by voltage (not radio, not time of day, not vars). A regulator may respond to an upstream or downstream capacitor switching, but that does not add up to many extra tap changes since the capacitor switches infrequently. Normally, the capacitor cannot cycle back and forth against the regulator. The only case might be if the regulator has negative settings for the reactive line-drop compensation.

With several regulators in series, adjustments to the time delay settings are the proper way to coordinate operations between units. Set the down-stream regulator with the longest time delay so it does not change taps excessively. For multiple regulators, increase the time delay with increasing distance from the source. Tap changes by a downstream regulator do not change the voltage upstream, but tap changes by an upstream regulator affect all downstream regulators. If a downstream regulator acts before the upstream regulator, the downstream regulator may have to tap again to meet its set voltage. Making the downstream regulator wait longer prevents it from tapping unnecessarily. Separate the time delays by at least 10 to 15 sec to allow the upstream unit to complete tap change operations.

5.5 Station Regulation

Utilities most commonly use load tap changing transformers (LTCs) to control distribution feeder voltages at the substation. In many cases (short, urban, thermally limited feeders) an LTC is all the voltage support a circuit needs.

An LTC or a stand-alone voltage regulator must compensate for the voltage change on the subtransmission circuit as well as the voltage drop through the transformer. Of these, the voltage drop through the transformer is normally the largest. Normally, the standard ±10% regulator can accomplish this. A regulator can hit the end of its range if the load has especially poor power factor. The voltage drop across a transformer follows:

$$V_{drop} = I_R \cdot R + I_X \cdot X$$

Since a transformer's X/R ratio is so high, the reactive portion of the load creates the most voltage drop across the transformer. Consider a 10% impedance transformer at full load with a load power factor of 0.8, which means the reactive power factor is 0.6. In this case, the voltage drop across the transformer is 6%. If the subtransmission voltage is 120 V (on a 120-V scale), the maximum that the regulator can boost the voltage to is 124 V. If this example had a transformer loaded to more than its base open-air rating (OA or ONAN), the regulator would be more limited in range. In most cases, we do not run into this problem as power factors are normally much better than these.

In most cases, bus regulation suffices. For cases where circuits have significant voltage drop, individual feeder regulation can be better. Individual feeder regulation also performs better on circuits with different load cycles. If commercial feeders are on the same bus as residential feeders, it is less likely that a bus regulator can keep voltages in line on all circuits. Normally, we handle this by using bus regulation and supplementary line regulators. In some cases, individual feeder regulation in the station is more appropriate.

The voltage on feeders serving secondary networks is controlled at the primary substation with LTC transformers. These circuits are short enough that feeder regulators are unnecessary. Network feeders are often supplied by parallel station transformers; paralleling LTC units raises several issues that are discussed in the next section.

5.5.1 Parallel Operation

With care, we can parallel regulators. The most common situation is in a substation where a utility wants to parallel two LTC transformers. If two paralleled transformers do not have the same turns ratio, current will circulate to balance the voltages. The circulating current is purely reactive, but it adds extra loading on the transformer.

Some of the methods to operate LTC transformers in parallel (Jauch, 2001; Westinghouse Electric Corporation, 1965) include

- *Negative-reactance control* — The reactance setting in the line-drop compensator is set to a negative value, so higher reactive current forces the control to lower taps. The transformer with the higher tap has more reactive current, and the transformer with the lower tap is absorbing this reactive current (it looks capacitive to this transformer). So, a negative-reactance setting forces the transformer with the highest tap (and most reactive current) to lower its taps and bring it into alignment with the other unit. This method limits the use of line-drop compensation and can lead to lower bus voltages.

- *Master-follower* — One controller, the master, regulates the voltage and signals the other tap changers (the followers) to match the tap setting. The master control normally gets feedback from the followers to confirm their operation.

- *Var balancing* — The controller adjusts taps as required to equalize the vars flowing in parallel transformers. Auxiliary circuitry is required. This method has the advantage that it works with transformers fed from separate transmission sources.

- *Circulating current method* — This is the most common control. Auxiliary circuitry is added to separate the load current through each transformer from the circulating current. Each transformer LTC control is fed the load current. The controller adjusts taps to minimize the difference in current between parallel units. Removing a unit does not require changing controller settings.

The complications associated with paralleling regulators are another reason utilities normally avoid closed bus ties in distribution substations.

5.5.2 Bus Regulation Settings

Although too often left unused, bus regulators (whether stand-alone regulators or load tap changing transformers) can use line-drop compensation. The concept of a load center rarely has good meaning for a bus supporting several circuits, but the voltage spread methods allow the regulator to boost voltage under heavy load.

The voltage-spread equations assume that the power factor at full load is the same as the power factor at light load. If the power factor is different at light and peak loads, we can use this information to provide more precise settings. We could solve the following to find new R and X settings with different power factors

$$V_{max} - V_{min} = (pf_{max} \cdot R_{set} + qf_{max} \cdot X_{set})I_{max} - (pf_{min} \cdot R_{set} + qf_{min} \cdot X_{set})I_{min}$$

However, it is easier to use the equations in Section 5.4.1.2 and use the average of the power factor at peak load and the power factor at light load. With line-drop compensation for bus regulation, the voltage-override feature helps to ensure that the LTC or regulator does not cause excessive voltages.

Individual substation feeder regulators are set the same as line feeder regulators. We can tune controller settings more precisely based on the individual characteristics of a given feeder. If the first part of the feeder is an express feeder with no load on it, we could boost the voltage higher than normal, especially if the circuit is voltage limited. Our main constraint is making sure that the first customer does not have high voltage.

5.6 Line Loss and Voltage Drop Relationships

Line losses are from the line current flowing through the resistance of the conductors. After distribution transformer losses, primary line losses are the largest cause of losses on the distribution system. Like any resistive losses, line losses are a function of the current squared multiplied by the resistance (I^2R). Ways to reduce line losses include

- Use a higher system voltage
- Balance circuits

- Convert single-phase circuits to three-phase circuits
- Reduce loads
- Increase power factor (capacitors)
- Reconductor with a larger size

Because losses are a function of the current squared, most losses occur on the primary near the substation. Losses occur regardless of the power factor of the circuit. Reducing the reactive portion of current reduces the total current, which can significantly impact losses.

Approximations using uniform load distributions are useful. A uniformly distributed load along a circuit of length l has the same losses as a single lumped load placed at a length of $l/3$ from the source end. For voltage drop, the equivalent circuits are different: a uniformly distributed load along a circuit of length l has the same voltage drop as a single lumped load placed at a length of $l/2$ from the source end. This $1/2$ rule for voltage drop and the $1/3$ rule for losses are helpful approximations when doing hand calculations or when making simplifications to enter in a load-flow program.

For a uniformly increasing load, the equivalent lumped load is at $0.53l$ of the length from the source. Figure 5.9 shows equivalent circuits for a uniform load and a uniformly increasing load.

Line losses decrease as operating voltage increases because the current decreases. Schultz (1978) derived several expressions for primary feeder I^2R losses on circuits with uniform load densities. His analysis showed that most 15 to 35 kV circuits are not voltage-drop limited — most are thermally limited. As the system voltage varies, the losses change the most for voltage-limited circuits (Schultz, 1978):

$$L_2 = \left(\frac{V_1}{V_2}\right)^2 L_1 \qquad \text{for a voltage-limited circuit}$$

$$L_2 = \left(\frac{V_1}{V_2}\right)^{2/3} L_1 \qquad \text{for a thermally-limited circuit}$$

where
V_1, V_2 = voltage on circuits 1 and 2
L_1, L_2 = feeder I^2R losses on circuits 1 and 2

On a system-wide basis, losses are expected to change with voltage with an exponent somewhere between $2/3$ and 2.

Losses, voltage drop, and capacity are all interrelated. Three-phase circuits have the highest power transfer capacity, the lowest voltage drop, and the lowest losses. Table 5.7 compares capacity, voltage drop, and losses of a balanced three-phase system with several other phasing configurations.

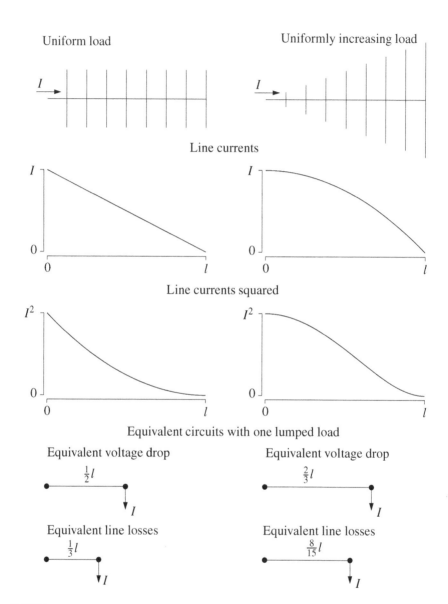

FIGURE 5.9
Equivalent circuits of uniform loads.

TABLE 5.7

Characteristics of Various Systems

System	Capacity in per Unit	Voltage Drop in per Unit for Equal kVA	Line Losses in per Unit for Equal kVA
Balanced three phase	1.0	1.0	1.0
Two phases	0.5	2.0	2.0
Two phases and a multigrounded neutral	0.67	2.0–3.3	1.2–3.0
Two phases and a unigrounded neutral	0.67	2.5–4.5	2.25
One phase and a multigrounded neutral	0.33	3.7–4.5	3.5–4.0
One phase and a unigrounded neutral	0.33	6.0	6.0

Note: The two-phase circuits assume all load is connected line to ground. Neutrals are the same size as the phases. Reduced neutrals increase voltage drop and (usually) line losses. The voltage drop and line loss ratios for circuits with multigrounded neutrals vary with conductor size.

Utilities consider both peak losses and energy losses. Peak losses are important because they compose a portion of the peak demand; energy losses are the total kilowatt-hours wasted as heat in the conductors. The peak losses are more easily estimated from measurements and models. The average losses can be found from the peak losses using the loss factor F_{ls}:

$$F_{ls} = \frac{\text{Average losses}}{\text{Peak losses}}$$

Normally, we do not have enough information to directly measure the loss factor. We do have the load factor (the average demand over the peak demand). The loss factor is some function of the load factor squared. The most common approximation (Gangel and Propst, 1965) is

$$F_{ls} = 0.15F_{ld} + 0.85F_{ld}^2$$

This is often used for evaluating line losses and transformer load losses (which are also a function of I^2R). Load factors closer to one result in loss factors closer to one. Another common expression is $F_{ls} = 0.3F_{ld} + 0.7F_{ld}^2$. Figure 5.10 shows both relationships.

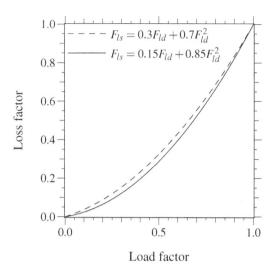

FIGURE 5.10
Relationship between load factor and loss factor.

References

ANSI C84.1-1995, *American National Standards for Electric Power Systems and Equipment — Voltage Ratings (60 Hz)*.

Beckwith, *Basic Considerations for the Application of LTC Transformers and Associated Controls*, Beckwith Electric Company, Application Note #17, 1998.

Bishop, M. T., Foster, J. D., and Down, D. A., "The Application of Single-Phase Voltage Regulators on Three-Phase Distribution Systems," *IEEE Industry Applications Magazine*, pp. 38–44, July/August 1996.

Brice, C. W., "Comparison of Approximate and Exact Voltage Drop Calculations for Distribution Lines," *IEEE Transactions on Power Apparatus and Systems*, vol. PAS-101, no. 11, pp. 4428–31, November, 1982.

Cooper Power Systems, "Determination of Regulator Compensator Settings," 1978. Publication R225-10-1.

Gangel, M. W. and Propst, R. F., "Distribution Transformer Load Characteristics," *IEEE Transactions on Power Apparatus and Systems*, vol. 84, pp. 671–84, August 1965.

General Electric, *Omnitext*, 1979. GET-3537B.

IEEE Std. C57.15-1999, *IEEE Standard Requirements, Terminology, and Test Code for Step-Voltage Regulators*.

Jauch, E. T., "Advanced Transformer Paralleling," IEEE/PES Transmission and Distribution Conference and Exposition, 2001.

Kirshner, D., "Implementation of Conservation Voltage Reduction at Commonwealth Edison," *IEEE Transactions on Power Systems*, vol. 5, no. 4, pp. 1178–82, November 1990.

Kirshner, D. and Giorsetto, P., "Statistical Tests of Energy Savings Due to Voltage Reduction," *IEEE Transactions on Power Apparatus and Systems*, vol. PAS-103, no. 6, pp. 1205–10, June 1984.

Lokay, H. E. and Custard, R. L., "A Field Method for Determining the Leading and Lagging Regulator in an Open-Delta Connection," *AIEE Transactions*, vol. 73, Part III, pp. 1684–6, 1954.

McCarthy, C., "CAPS — Choosing the Feeders, Part I," in *Systems Engineering Technical Update*: Cooper Power Systems, 2000.

Priess, R. F. and Warnock, V. J., "Impact of Voltage Reduction on Energy and Demand," *IEEE Transactions on Power Apparatus and Systems*, vol. PAS-97, no. 5, pp. 1665–71, Sept/Oct 1978.

Schultz, N. R., "Distribution Primary Feeder I²R Losses," *IEEE Transactions on Power Apparatus and Systems*, vol. PAS-97, no. 2, pp. 603–9, March–April 1978.

Sen, P. K. and Larson, S. L., "Fundamental Concepts of Regulating Distribution System Voltages," IEEE Rural Electric Power Conference, Department of Electrical Engineering, Colorado University, Denver, CO, 1994.

Westinghouse Electric Corporation, *Distribution Systems*, vol. 3, 1965.

Willis, H. L., "Characteristics of Distribution Loads," in *Electrical Transmission and Distribution Reference Book*. Raleigh, NC: ABB Power T&D Company, 1997a.

Willis, H. L., *Power Distribution Planning Reference Book*, Marcel Dekker, New York, 1997b.

Regs? Treat them with respect. They are a transformer. Anyone who has dropped load with a tx knows that you can build a fire if you don't take the load into consideration. The difference with regs is that the load is the feeder. Get it?

anonymous post
www.powerlineman.com

6

Capacitor Application

Capacitors provide tremendous benefits to distribution system performance. Most noticeably, capacitors reduce losses, free up capacity, and reduce voltage drop:

- *Losses; Capacity* — By canceling the reactive power to motors and other loads with low power factor, capacitors decrease the line current. Reduced current frees up capacity; the same circuit can serve more load. Reduced current also significantly lowers the I^2R line losses.

- *Voltage drop* — Capacitors provide a voltage boost, which cancels part of the drop caused by system loads. Switched capacitors can regulate voltage on a circuit.

If applied properly and controlled, capacitors can significantly improve the performance of distribution circuits. But if not properly applied or controlled, the reactive power from capacitor banks can create losses and high voltages. The greatest danger of overvoltages occurs under light load. Good planning helps ensure that capacitors are sited properly. More sophisticated controllers (like two-way radios with monitoring) reduce the risk of improperly controlling capacitors, compared to simple controllers (like a time clock).

Capacitors work their magic by storing energy. Capacitors are simple devices: two metal plates sandwiched around an insulating dielectric. When charged to a given voltage, opposing charges fill the plates on either side of the dielectric. The strong attraction of the charges across the very short distance separating them makes a tank of energy. Capacitors oppose changes in voltage; it takes time to fill up the plates with charge, and once charged, it takes time to discharge the voltage.

On ac power systems, capacitors do not store their energy very long — just one-half cycle. Each half cycle, a capacitor charges up and then discharges its stored energy back into the system. The net real power transfer is zero. Capacitors provide power just when reactive loads need it. Just when a motor with low power factor needs power from the system, the capacitor is there to provide it. Then in the next half cycle, the motor releases its excess energy, and the capacitor is there to absorb it. Capacitors and reactive loads

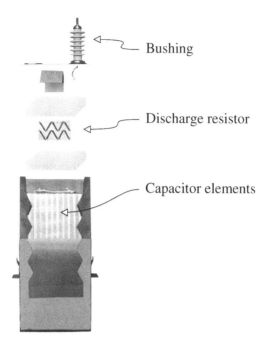

Bushing

Discharge resistor

Capacitor elements

FIGURE 6.1
Capacitor components. (From General Electric Company. With permission.)

exchange this reactive power back and forth. This benefits the system because that reactive power (and extra current) does not have to be transmitted from the generators all the way through many transformers and many miles of lines; the capacitors can provide the reactive power locally. This frees up the lines to carry real power, power that actually does work.

Capacitor units are made of series and parallel combinations of capacitor packs or elements put together as shown in Figure 6.1. Capacitor elements have sheets of polypropylene film, less than one mil thick, sandwiched between aluminum foil sheets. Capacitor dielectrics must withstand on the order of 2000 V/mil (78 kV/mm). No other medium-voltage equipment has such high voltage stress. An underground cable for a 12.47-kV system has insulation that is at least 0.175 in. (4.4 mm) thick. A capacitor on the same system has an insulation separation of only 0.004 in. (0.1 mm).

Utilities often install substation capacitors and capacitors at points on distribution feeders. Most feeder capacitor banks are pole mounted, the least expensive way to install distribution capacitors. Pole-mounted capacitors normally provide 300 to 3600 kvar at each installation. Many capacitors are switched, either based on a local controller or from a centralized controller through a communication medium. A line capacitor installation has the capacitor units as well as other components, possibly including arresters, fuses, a control power transformer, switches, and a controller (see Figure 6.2 for an example).

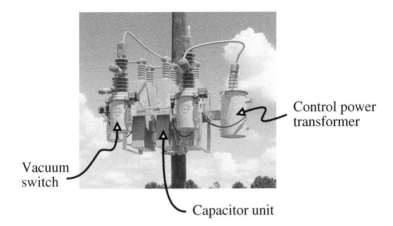

Control power transformer

Vacuum switch

Capacitor unit

FIGURE 6.2
Overhead line capacitor installation. (From Cooper Power Systems, Inc. With permission.)

While most capacitors are pole mounted, some manufacturers provide padmounted capacitors. As more circuits are put underground, the need for padmounted capacitors will grow. Padmounted capacitors contain capacitor cans, switches, and fusing in a deadfront package following standard padmounted-enclosure integrity requirements (ANSI C57.12.28-1998). These units are much larger than padmounted transformers, so they must be sited more carefully to avoid complaints due to aesthetics. The biggest obstacles are cost and aesthetics. The main complaint is that padmounted capacitors are large. Customers complain about the intrusion and the aesthetics of such a large structure (see Figure 6.3).

FIGURE 6.3
Example padmounted capacitor. (From Northeast Power Systems, Inc. With permission.)

TABLE 6.1

Substation vs. Feeder Capacitors

Advantages	Disadvantages
Feeder Capacitors	
Reduces line losses	More difficult to control reliably
Reduces voltage drop along the feeder	Size and placement important
Frees up feeder capacity	
Lower cost	
Substation Capacitors	
Better control	No reduction in line losses
Best placement if leading vars are needed for system voltage support	No reduction in feeder voltage drop
	Higher cost

Substation capacitors are normally offered as open-air racks. Normally elevated to reduce the hazard, individual capacitor units are stacked in rows to provide large quantities of reactive power. All equipment is exposed. Stack racks require a large substation footprint and are normally engineered for the given substation. Manufacturers also offer metal-enclosed capacitors, where capacitors, switches, and fuses (normally current-limiting) are all enclosed in a metal housing.

Substation capacitors and feeder capacitors both have their uses. Feeder capacitors are closer to the loads — capacitors closer to loads more effectively release capacity, improve voltage profiles, and reduce line losses. This is especially true on long feeders that have considerable line losses and voltage drop. Table 6.1 highlights some of the differences between feeder and station capacitors. Substation capacitors are better when more precise control is needed. System operators can easily control substation capacitors wired into a SCADA system to dispatch vars as needed. Modern communication and control technologies applied to feeder capacitors have reduced this advantage. Operators can control feeder banks with communications just like station banks, although some utilities have found the reliability of switched feeder banks to be less than desired, and the best times for switching in vars needed by the system may not correspond to the best time to switch the capacitor in for the circuit it is located on.

Substation capacitors may also be desirable if a leading power factor is needed for voltage support. If the power factor is leading, moving this capacitor out on the feeder increases losses. Substation capacitors cost more than feeder capacitors. This may seem surprising, but we must individually engineer station capacitors, and the space they take up in a station is often valuable real estate. Pole-mounted capacitor installations are more standardized.

Utilities normally apply capacitors on three-phase sections. Applications on single-phase lines are done but less common. Application of three-phase banks downstream of single-phase protectors is normally not done because of ferroresonance concerns. Most three-phase banks are connected grounded-wye on four-wire multigrounded circuits. Some are connected in floating wye. On three-wire circuits, banks are normally connected as a floating wye.

Most utilities also include arresters and fuses on capacitor installations. Arresters protect capacitor banks from lightning-overvoltages. Fuses isolate failed capacitor units from the system and clear the fault before the capacitor fails violently. In high fault-current areas, utilities may use current-limiting fuses. Switched capacitor units normally have oil or vacuum switches in addition to a controller. Depending on the type of control, the installation may include a control power transformer for power and voltage sensing and possibly a current sensor. Because a capacitor bank has a number of components, capacitors normally are not applied on poles with other equipment.

Properly applied capacitors return their investment very quickly. Capacitors save significant amounts of money in reduced losses. In some cases, reduced loadings and extra capacity can also delay building more distribution infrastructure.

6.1 Capacitor Ratings

Capacitor units rated from 50 to over 500 kvar are available; Table 6.2 shows common capacitor unit ratings. A capacitor's rated kvar is the kvar at rated voltage. Three-phase capacitor banks are normally referred to by the total kvar on all three phases. Distribution feeder banks normally have one or two or (more rarely) three units per phase. Many common size banks only have one capacitor unit per phase.

IEEE Std. 18 defines standards for capacitors and provides application guidelines. Capacitors should not be applied when any of the following limits are exceeded (IEEE Std. 18-2002):

- 135% of nameplate kvar
- 110% of rated rms voltage, and crest voltage not exceeding $1.2\sqrt{2}$ of rated rms voltage, including harmonics but excluding transients
- 135% of nominal rms current based on rated kvar and rated voltage

Capacitor dielectrics must withstand high voltage stresses during normal operation — on the order of 2000 V/mil. Capacitors are designed to withstand overvoltages for short periods of time. IEEE Std. 18-1992 allows up to 300 power-frequency overvoltages within the time durations in Table 6.3 (without transients or harmonic content). New capacitors are tested with at least a 10-sec overvoltage, either a dc-test voltage of 4.3 times rated rms or an ac voltage of twice the rated rms voltage (IEEE Std. 18-2002).

TABLE 6.2

Common Capacitor Unit Ratings

Volts, rms (Terminal-to-Terminal)	kvar	Number of Phases	BIL, kV
216	5, 7 1/2, 13 1/3, 20, and 25	1 and 3	30
240	2.5, 5, 7 1/2, 10, 15, 20, 25, and 50	1 and 3	30
480, 600	5, 10, 15, 20, 25, 35, 50, 60, and 100	1 and 3	30
2400	50, 100, 150, 200, 300, and 400	1 and 3	75, 95, 125, 150, and 200
2770	50, 100, 150, 200, 300, 400, and 500	1 and 3	75, 95, 125, 150, and 200
4160, 4800	50, 100, 150, 200, 300, 400, 500, 600, 700, and 800	1 and 3	75, 95, 125, 150, and 200
6640, 7200, 7620, 7960, 8320, 9540, 9960, 11,400, 12,470, 13,280, 13,800, 14,400	50, 100, 150, 200, 300, 400, 500, 600, 700, and 800	1	95, 125, 150, and 200
15,125	50, 100, 150, 200, 300, 400, 500, 600, 700, and 800	1	125, 150, and 200
19,920	100, 150, 200, 300, 400, 500, 600, 700, and 800	1	125, 150, and 200
20,800, 21,600, 22,800, 23,800, 24,940	100, 150, 200, 300, 400, 500, 600, 700, and 800	1	150 and 200

Source: IEEE Std. 18-2002. Copyright 2002 IEEE. All rights reserved.

TABLE 6.3

Maximum Permissible Power-Frequency Voltages

Duration	Maximum Permissible Voltage (multiplying factor to be applied to rated voltage rms)
6 cycles	2.20
15 cycles	2.00
1 sec	1.70
15 sec	1.40
1 min	1.30
30 min	1.25
Continuous	1.10

Note: This is not in IEEE Std. 18-2002 but it will be addressed in IEEE's updated capacitor application guide.

Source: ANSI/IEEE Std. 18-1992. Copyright 1993 IEEE. All rights reserved.

Capacitors should withstand various peak voltage and current transients; the allowable peak depends on the number of transients expected per year (see Table 6.4).

The capacitance of a unit in microfarads is

TABLE 6.4

Expected Transient Overcurrent and Overvoltage Capability

Probable Number of Transients per year	Permissible Peak Transient Current (multiplying factor to be applied to rated rms current)	Permissible Peak Transient Voltage (multiplying factor to be applied to rated rms voltage)
4	1500	5.0
40	1150	4.0
400	800	3.4
4000	400	2.9

Note: This is not in IEEE Std. 18-2002, but it will be addressed in IEEE's updated capacitor application guide.

Source: ANSI/IEEE Std. 18-1992. Copyright 1993 IEEE. All rights reserved.

$$C_{uF} = \frac{2.65 Q_{kvar}}{V_{kV}^2}$$

where

V_{kV} = capacitor voltage rating, kV

Q_{kvar} = unit reactive power rating, kvar

Capacitors are made within a given tolerance. The IEEE standard allows reactive power to range between 100 and 110% when applied at rated sinusoidal voltage and frequency (at 25°C case and internal temperature) (IEEE Std. 18-2002). Older units were allowed to range up to 115% (ANSI/IEEE Std. 18-1992). Therefore, the capacitance also must be between 100 and 110% of the value calculated at rated kvar and voltage. In practice, most units are from +0.5 to +4.0%, and a given batch is normally very uniform.

Capacitor losses are typically on the order of 0.07 to 0.15 W/kvar at nominal frequency. Losses include resistive losses in the foil, dielectric losses, and losses in the internal discharge resistor.

Capacitors must have an internal resistor that discharges a capacitor to 50 V or less within 5 min when the capacitor is charged to the peak of its rated voltage $(\sqrt{2}V_{rms})$. This resistor is the major component of losses within a capacitor. The resistor must be low enough such that the RC time constant causes it to decay in 300 sec as

$$\frac{50}{\sqrt{2V}} \le e^{-300/RC}$$

where

V = capacitor voltage rating, V

R = discharge resistance, Ω

C = capacitance, F

TABLE 6.5

Maximum Ambient Temperatures for Capacitor Application

Mounting Arrangement	Ambient Air Temperature (°C) 4-h Average[a]
Isolated capacitor	46
Single row of capacitors	46
Multiple rows and tiers of capacitors	40
Metal-enclosed or -housed equipments	40

[a] The arithmetic average of the four consecutive highest hourly readings during the hottest day expected at that location.

Source: IEEE Std. 18-2002. Copyright 2002 IEEE. All rights reserved.

So, the discharge resistor must continually dissipate at least the following power in watts:

$$P_{watts} = -\frac{Q_{kvar}}{113.2} \ln\left(\frac{35.36}{V}\right)$$

where Q_{kvar} is the capacitor rating (single or three phase). For 7.2-kV capacitors, the lower bound on losses is 0.047 W/kvar.

Some utilities use a shorting bar across the terminals of capacitors during shipping and in storage. The standard recommends waiting for 5 min to allow the capacitor to discharge through the internal resistor.

Capacitors have very low losses, so they run very cool. But capacitors are very sensitive to temperature and are rated for lower temperatures than other power system equipment such as cables or transformers. Capacitors do not have load cycles like transformers; they are always at full load. Also, capacitors are designed to operate at high dielectric stresses, so they have less margin for degraded insulation. Standards specify an upper limit for application of 40 or 46°C depending on arrangement (see Table 6.5). These limits assume unrestricted ventilation and direct sunlight. At the lower end, IEEE standard 18 specifies that capacitors shall be able to operate continuously in a −40°C ambient.

6.2 Released Capacity

In addition to reducing losses and improving voltage, capacitors release capacity. Improving the power factor increases the amount of real-power load the circuit can supply. Using capacitors to supply reactive power reduces the amount of current in the line, so a line of a given ampacity can

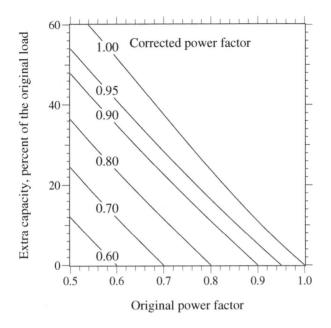

FIGURE 6.4
Released capacity with improved power factor.

carry more load. Figure 6.4 shows that capacitors release significant capacity, especially if the original power factor is low. Figure 6.5 shows another way to view the extra capacity, as a function of the size of capacitor added.

6.3 Voltage Support

Capacitors are constant-impedance devices. At higher voltages, capacitors draw more current and produce more reactive power as

$$I = I_{rated}V_{pu} \quad \text{and} \quad Q_{kvar} = Q_{rated}V_{pu}{}^2$$

where V_{pu} is the voltage in per unit of the capacitor's voltage rating. Capacitors applied at voltages other than their rating provide vars in proportion to the per-unit voltage squared.

Capacitors provide almost a fixed voltage rise. The reactive current through the system impedance causes a voltage rise in percent of

$$V_{rise} = \frac{Q_{kvar}X_L}{10\,V_{kV,l\text{-}l}^2}$$

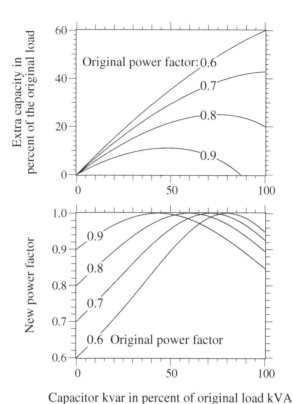

FIGURE 6.5
Extra capacity as a function of capacitor size.

where
X_L = positive-sequence system impedance from the source to the capacitor, Ω
$V_{kV,\,l-l}$ = line-to-line system voltage, kV
Q_{kvar} = three-phase bank rating, kvar

While this equation is very good for most applications, it is not exactly right because the capacitive current changes in proportion to voltage. At a higher operating voltage, a capacitor creates more voltage rise than the equation predicts.

Since the amount of voltage rise is dependent on the impedance upstream of the bank, to get the voltage boost along the entire circuit, put the capacitor at the end of the circuit. The best location for voltage support depends on where the voltage support is needed. Figure 6.6 shows how a capacitor changes the voltage profile along a circuit. Unlike a regulator, a capacitor changes the voltage profile upstream of the bank.

Table 6.6 shows the percentage voltage rise from capacitors for common conductors at different voltages. This table excludes the station transformer

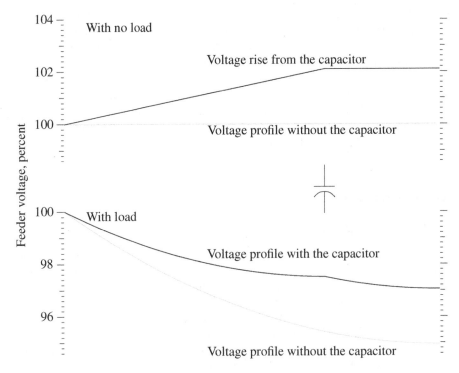

FIGURE 6.6
Voltage profiles after addition of a capacitor bank. (Copyright © 2002. Electric Power Research Institute. 1001691. *Improved Reliability of Switched Capacitor Banks and Capacitor Technology.* Reprinted with permission.)

TABLE 6.6

Percent Voltage Rise for Various Conductors and Voltage Levels

Conductor Size	X_L Ω/mi	Percent Voltage Rise per Mile with 100 kvar per Phase Line-to-Line System Voltage, kV			
		4.8	12.47	24.9	34.5
4	0.792	1.031	0.153	0.038	0.020
2	0.764	0.995	0.147	0.037	0.019
1/0	0.736	0.958	0.142	0.036	0.019
4/0	0.694	0.903	0.134	0.034	0.017
350	0.656	0.854	0.127	0.032	0.017
500	0.635	0.826	0.122	0.031	0.016
750	0.608	0.791	0.117	0.029	0.015

Note: Impedance are for all-aluminum conductors with GMD=4.8 ft.

impedance but still provides a useful approximation. Inductance does not change much with conductor size; the voltage change stays the same over a wide range of conductor sizes. For 15-kV class systems, capacitors increase the voltage by about 0.12% per mi per 100 kvar per phase.

On switched capacitor banks, the voltage change constrains the size of banks at some locations. Normally, utilities limit the voltage change to 3 to 4%. On a 12.47-kV circuit, a three-phase 1200-kvar bank boosts the voltage 4% at about 8 mi from the substation. To keep within a 4% limit, 1200-kvar banks must only be used within the first 8 mi of the station.

6.4   Reducing Line Losses

One of the main benefits of applying capacitors is that they can reduce distribution line losses. Losses come from current through the resistance of conductors. Some of that current transmits real power, but some flows to supply reactive power. Reactive power provides magnetizing for motors and other inductive loads. Reactive power does not spin kWh meters and performs no useful work, but it must be supplied. Using capacitors to supply reactive power reduces the amount of current in the line. Since line losses are a function of the current squared, I^2R, reducing reactive power flow on lines significantly reduces losses.

Engineers widely use the "2/3 rule" for sizing and placing capacitors to optimally reduce losses. Neagle and Samson (1956) developed a capacitor placement approach for uniformly distributed lines and showed that the optimal capacitor location is the point on the circuit where the reactive power flow equals half of the capacitor var rating. From this, they developed the 2/3 rule for selecting and placing capacitors. For a uniformly distributed load, the optimal size capacitor is 2/3 of the var requirements of the circuit. The optimal placement of this capacitor is 2/3 of the distance from the substation to the end of the line. For this optimal placement for a uniformly distributed load, the substation source provides vars for the first 1/3 of the circuit, and the capacitor provides vars for the last 2/3 of the circuit (see Figure 6.7).

A generalization of the 2/3 rule for applying n capacitors to a circuit is to size each one to $2/(2n+1)$ of the circuit var requirements. Apply them equally spaced, starting at a distance of $2/(2n+1)$ of the total line length from the substation and adding the rest of the units at intervals of $2/(2n+1)$ of the total line length. The total vars supplied by the capacitors is $2n/(2n+1)$ of the circuit's var requirements. So to apply three capacitors, size each to 2/7 of the total vars needed, and locate them at per unit distances of 2/7, 4/7, and 6/7 of the line length from the substation.

Grainger and Lee (1981) provide an optimal yet simple method for placing fixed capacitors on a circuit with any load profile, not just a uniformly

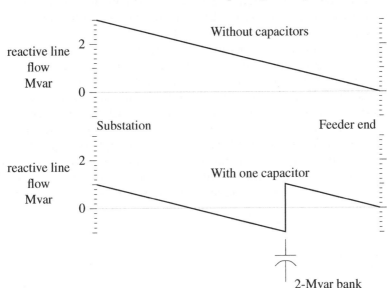

FIGURE 6.7
Optimal capacitor loss reduction using the two-thirds rule. (Copyright © 2002. Electric Power Research Institute. 1001691. *Improved Reliability of Switched Capacitor Banks and Capacitor Technology*. Reprinted with permission.)

distributed load. With the Grainger/Lee method, we use the reactive load profile of a circuit to place capacitors. The basic idea is again to locate banks at points on the circuit where the reactive power equals one half of the capacitor var rating. With this *1/2-kvar rule*, the capacitor supplies half of its vars downstream, and half are sent upstream. The basic steps of this approach are:

1. *Pick a size* — Choose a standard size capacitor. Common sizes range from 300 to 1200 kvar, with some sized up to 2400 kvar. If the bank size is 2/3 of the feeder requirement, we only need one bank. If the size is 1/6 of the feeder requirement, we need five capacitor banks.
2. *Locate the first bank* — Start from the end of the circuit. Locate the first bank at the point on the circuit where var flows on the line are equal to half of the capacitor var rating.
3. *Locate subsequent banks* — After a bank is placed, reevaluate the var profile. Move upstream until the next point where the var flow equals half of the capacitor rating. Continue placing banks in this manner until no more locations meet the criteria.

There is no reason we have to stick with the same size of banks. We could place a 300-kvar bank where the var flow equals 150 kvar, then apply a 600-

kvar bank where the var flow equals 300 kvar, and finally apply a 450-kvar bank where the var flow equals 225 kvar. Normally, it is more efficient to use standardized bank sizes, but different size banks at different portions of the feeder might help with voltage profiles.

The 1/2-kvar method works for any section of line. If a line has major branches, we can apply capacitors along the branches using the same method. Start at the end, move upstream, and apply capacitors at points where the line's kvar flow equals half of the kvar rating of the capacitor. It also works for lines that already have capacitors (it does not optimize the placement of all of the banks, but it optimizes placement of new banks). For large industrial loads, the best location is often going to be right at the load.

Figure 6.8 shows the optimal placement of 1200-kvar banks on an example circuit. Since the end of the circuit has reactive load above the 600-kvar threshold for sizing 1200-kvar banks, we apply the first capacitor at the end

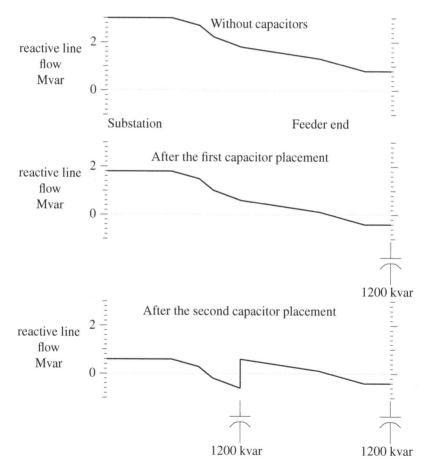

FIGURE 6.8
Placement of 1200-kvar banks using the 1/2-kvar method.

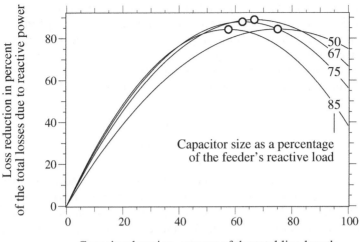

FIGURE 6.9
Sensitivity to losses of sizing and placing one capacitor on a circuit with a uniform load. (The circles mark the optimum location for each of the sizes shown.)

of the circuit. (The circuit at the end of the line could be one large customer or branches off the main line.) The second bank goes near the middle. The circuit has an express feeder near the start. Another 1200-kvar bank could go in just after the express feeder, but that does not buy us anything. The two capacitors total 2400 kvar, and the feeder load is 3000 kvar. We really need another 600-kvar capacitor to zero out the var flow before it gets to the express feeder.

Fortunately, capacitor placement and sizing does not have to be exact. Quite good loss reduction occurs even if sizing and placement are not exactly optimum. Figure 6.9 shows the loss reduction for one fixed capacitor on a circuit with a uniform load. The 2/3 rule specifies that the optimum distance is 2/3 of the distance from the substation and 2/3 of the circuit's var requirement. As long as the size and location are somewhat close (within 10%), the not-quite-optimal capacitor placement provides almost as much loss reduction as the optimal placement.

Consider the voltage impacts of capacitors. Under light load, check that the capacitors have not raised the voltages above allowable standards. If voltage limits are exceeded, reduce the size of the capacitor banks or the number of capacitor banks until voltage limits are not exceeded. If additional loss reduction is desired, consider switched banks as discussed below.

6.4.1 Energy Losses

Use the average reactive loading profile to optimally size and place capacitors for energy losses. If we use the peak-load case, the 1/2-kvar method optimizes losses during the peak load. If we have a load-flow case with the average reactive load, the 1/2-kvar method or the 2/3 rule optimizes energy losses. This leads to more separation between banks and less kvars applied than if we optimize for peak losses.

If an average system case is not available, then we can estimate it by scaling the peak load case by the reactive load factor, *RLF*:

$$RLF = \frac{\text{Average kvar Demand}}{\text{Peak kvar Demand}}$$

The reactive load factor is similar to the traditional load factor except that it only considers the reactive portion of the load. If we have no information on the reactive load factor, use the total load factor. Normally, the reactive load factor is higher than the total load factor. Figure 6.10 shows an example of power profiles; the real power (kW) fluctuates significantly more than the reactive power (kvar).

6.5   Switched Banks

Switched banks provide benefits under the following situations:

- *More loss reduction* — As the reactive loading on the circuit changes, we reduce losses by switching banks on and off to track these changes.
- *Voltage limits* — If optimally applied banks under the average loading scenario cause excessive voltage under light load, then use switched banks.

In addition, automated capacitors — those with communications — have the flexibility to also use distribution vars for transmission support.

Fixed banks are relatively easy to site and size optimally. Switched banks are more difficult. Optimally sizing capacitors, placing them, and deciding when to switch them are difficult tasks. Several software packages are available that can optimize this solution. This is an intensely studied area, and technical literature documents several approaches (among these Carlisle and El-Keib, 2000; Grainger and Civanlar, 1985; Shyh, 2000).

To place switched capacitors using the 1/2-kvar method, again place the banks at the location where the line kvar equals half the bank rating. But instead of using the average reactive load profile (the rule for fixed banks), use the average reactive flow during the time the capacitor is on. With time-switched banks and information on load profiles (or typical load profiles),

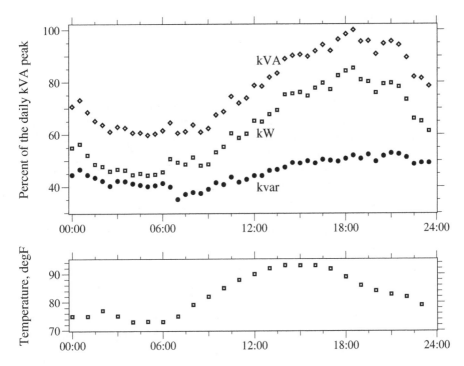

FIGURE 6.10
Example of real and reactive power profiles on a residential feeder on a peak summer day with 95% air conditioning. (Data from East Central Oklahoma Electric Cooperative, Inc. [RUS 1724D-112, 2001].)

we can pick the on time and the off time and determine the proper sizing based on the average reactive flow between the on and off times. Or, we can place a bank and pick the on and off times such that the average reactive line flow while the bank is switched on equals half of the bank rating. In these cases, we have specified the size and either the placement or switching time. To more generally optimize — including sizing, placement, number of banks, and switching time — we must use a computer, which iterates to find a solution [see Lee and Grainger (1981) for one example].

Combinations of fixed and switched banks are more difficult. The following approach is not optimal but gives reasonable results. Apply fixed banks to the circuit with the 1/2-kvar rule based on the light-load case. Check voltages. If there are undervoltages, increase the size of capacitors, use more capacitor banks, or add regulators. Now, look for locations suitable for switched banks. Again, use the average reactive line flows for the time when the capacitor is on (with the already-placed fixed capacitors in the circuit model). When applying switched capacitors, check the light-load case for possible overvoltages, and check the peak-load case for under-voltages.

6.6 Local Controls

Several options for controls are available for capacitor banks:

- *Time clock* — The simplest scheme: the controller switches capacitors on and off based on the time of day. The on time and the off time are programmable. Modern controllers allow settings for weekends and holidays. This control is the cheapest but also the most susceptible to energizing the capacitor at the wrong time (due to loads being different from those expected, to holidays or other unexpected light periods, and especially to mistakenly set or inaccurate clocks). Time clock control is predictable; capacitors switch on and off at known times and the controller limits the number of switching operations (one energization and one deenergization per day).

- *Temperature* — Another simple control; the controller switches the capacitor bank on or off depending on temperature. Normally these might be set to turn the capacitors on in the range of 85 and 90°F and turn them off at temperatures somewhere between 75 and 80°F.

- *Voltage* — The capacitor switches on and off, based on voltage. The user provides the threshold minimum and maximum voltages as well as time delays and bandwidths to prevent excessive operations. Voltage control is most appropriate when the primary role of a capacitor is voltage support and regulation.

- *Vars* — The capacitor uses var measurements to determine switching. This is the most accurate method of ensuring that the capacitor is on at the appropriate times for maximum reduction of losses.

- *Power factor* — Similar to var control, the controller switches capacitors on and off based on the measured power factor. This is rarely used by utilities.

- *Current* — The capacitor switches on and off based on the line current (as measured downstream of the capacitor). While not as effective as var control, current control does engage the capacitor during heavy loads, which usually corresponds to the highest needs for vars.

Many controllers offer many or all of these possibilities. Many are usable in combination; turn capacitors on for low voltage or for high temperature.

Var, power factor, voltage, or current controllers require voltage or current sensing or both. To minimize cost and complexity, controllers often switch all three phases using sensors on just one phase. A control power transformer is often also used to sense voltage. While unusual, Alabama Power switches each phase independently depending on the var requirements of each phase (Clark, 2001); this optimizes loss reduction and helps reduce unbalance. Because capacitor structures are rather busy, some utilities like to use voltage and/or current-sensing insulators. Meter-grade accuracy is not needed for controlling capacitors.

To coordinate more than one capacitor with switched var controls, set the most-distant unit to have the shortest time delay. Increase the time delay on successive units progressing back to the substation. This leaves the unit closest to the substation with the longest time delay. The most distant unit switches first. Upstream units see the change and do not need to respond. This strategy is the opposite of that used for coordinating multiple line voltage regulators.

For var-controlled banks, locate the current sensor on the source (substation) side of the bank. Then, the controller can detect the reactive power change when the capacitor switches. To properly calculate vars, the wiring for the CT and PT must provide correct polarities to the controller.

One manufacturer provides the following rules of thumb for setting var control trip and close settings (Fisher Pierce, 2000):

- Close setpoint: 2/3 × capacitor bank size (in kvar), lagging.
- Trip setpoint: Close set point – 1.25 × bank size, will be leading. (This assumes that the CT is on the source side of the bank.)

For a 600-kvar bank application, this yields

Close setpoint: 2/3 × 600 = +400 kvar (lagging)

Trip setpoint: 400 – 1.25 × 600 = –350 kvar (leading)

For this example, the unit trips when the load kvar drops below +250 kvar (lagging). This effectively gives a bandwidth wide enough (+400 to +250 kvar) to prevent excessive switching operations in most cases.

Voltage-controlled capacitor banks have bandwidths. Normally, we want the bandwidth to be at least 1.5 times the expected voltage change due to the capacitor bank. Ensure that the bandwidth is at least 3 or 4 V (on a 120-V scale). Set the trip setting below the normal light-load voltage (or the bank will never switch off).

If a switched capacitor is located on a circuit that can be operated from either direction, make sure the controller mode can handle operation with power flow in either direction. Time-of-day, temperature, and voltage control are not affected by reverse power flow; var, current, and power factor control are affected. Some controllers can sense reverse power and shift control modes. One model provides several options if it detects reverse power: switch to voltage mode, calculate var control while accounting for the effect of the capacitor bank, inhibit switching, trip and lock out the bank, or close and hold the bank in. If a circuit has distributed generation, we do not want to shift modes based on reverse power flow; the controller should shift modes only for a change in direction to the system source.

Capacitor controllers normally have counters to record the number of operations. The counters help identify when to perform maintenance and can identify control-setting problems. For installations that are excessively switching, modify control settings, time delays, or bandwidths to reduce switching. Some controllers can limit the number of switch operations within a given time period to reduce wear on capacitor switches.

Voltage control provides extra safety to prevent capacitors from causing overvoltages. Some controllers offer types of voltage override control; the primary control may be current, vars, temperature, or time of day, but the controller trips the bank if it detects excessive voltage. A controller may also restrain from switching in if the extra voltage rise from the bank would push the voltage above a given limit.

6.7   Automated Controls

Riding the tide of lower-cost wireless communication technologies, many utilities have automated capacitor banks. Many of the cost reductions and feature improvements in communication systems have resulted from the proliferation of cellular phones, pagers, and other wireless technologies used by consumers and by industry. Controlling capacitors requires little band-width, so high-speed connections are unnecessary. This technology changes quickly. The most common communications systems for distribution line capacitors are 900-MHz radio systems, pager systems, cellular phone systems, cellular telemetric systems, and VHF radio. Some of the common features of each are

- *900-MHz radio* — Very common. Several spread-spectrum data radios are available that cover 902–928 MHz applications. A private network requires an infrastructure of transmission towers.

- *Pager systems* — Mostly one-way, but some two-way, communications. Pagers offer inexpensive communications options, especially for infrequent usage. One-way communication coverage is wide-spread; two-way coverage is more limited (clustered around major cities). Many of the commercial paging networks are suitable for capacitor switching applications.

- *Cellular phone systems* — These use one of the cellular networks to provide two-way communications. Many vendors offer cellular modems for use with several cellular networks. Coverage is typically very good.

- *Cellular telemetric systems* — These use the unused data component of cellular signals that are licensed on existing cellular networks. They allow only very small messages to be sent — enough, though, to perform basic capacitor automation needs. Coverage is typically very good, the same as regular cellular coverage.

- *VHF radio* — Inexpensive one-way communications are possible with VHF radio communication. VHF radio bands are available for telemetry uses such as this. Another option is a simulcast FM signal that uses extra bandwidth available in the commercial FM band.

Standard communication protocols help ease the building of automated infrastructures. Equipment and databases are more easily interfaced with standard protocols. Common communication protocols used today for SCADA applications and utility control systems include DNP3, IEC 870, and Modbus.

DNP 3.0 (Distributed Network Protocol) is the most widely used standard protocol for capacitor controllers (DNP Users Group, 2000). It originated in the electric industry in America with Harris Distributed Automation Products and was based on drafts of the IEC870-5 SCADA protocol standards (now known as IEC 60870-5). DNP supports master–slave and peer-to-peer communication architectures. The protocol allows extensions while still providing interoperability. Data objects can be added to the protocol without affecting the way that devices interoperate. DNP3 was designed for transmitting data acquisition information and control commands from one computer to another. (It is not a general purpose protocol for hypertext, multimedia, or huge files.)

One-way or two-way — we can remotely control capacitors either way. Two-way communication has several advantages:

- *Feedback* — A local controller can confirm that a capacitor switched on or off successfully. Utilities can use the feedback from two-way communications to dispatch crews to fix capacitor banks with blown fuses, stuck switches, misoperating controllers, or other problems.

- *Voltage/var information* — Local information on line var flows and line voltages allows the control to more optimally switch capacitor banks to reduce losses and keep voltages within limits.

- *Load flows* — Voltage, current, and power flow information from pole-mounted capacitor banks can be used to update and verify load-flow models of a system. The information can also help when tracking down customer voltage, stray voltage, or other power quality problems. Loading data helps utilities monitor load growth and plan for future upgrades. One utility even uses capacitor controllers to capture fault location information helping crews to locate faults.

When a controller only has one-way communications, a local voltage override control feature is often used. The controller blocks energizing a capacitor bank if doing so would push the voltage over limits set by the user.

Several schemes and combinations of schemes are used to control capacitors remotely:

- *Operator dispatch* — Most schemes allow operators to dispatch distribution capacitors. This feature is one of the key reasons utilities automate capacitor banks. Operators can dispatch distribution capacitors just like large station banks. If vars are needed for transmission support, large numbers of distribution banks can be switched on. This control scheme is usually used in conjunction with other controls.

- *Time scheduling* — Capacitors can be remotely switched, based on the time of day and possibly the season or temperature. While this may seem like an expensive time control, it still allows operators to override the schedule and dispatch vars as needed.

- *Substation var measurement* — A common way to control feeder capacitors is to dispatch based on var/power factor measurements in the substation. If a feeder has three capacitor banks, they are switched on or off in some specified order based on the power factor on the feeder measured in the substation.

- *Others* — More advanced (and complicated) algorithms can dispatch capacitors based on a combination of local var measurements and voltage measurements along with substation var measurements.

6.8   Reliability

Several problems contribute to the overall reliability or unreliability of capacitor banks. In a detailed analysis of Kansas City Power & Light's automated capacitor banks, Goeckeler (1999) reported that blown fuses are KCP&L's biggest problem, but several other problems exist (Table 6.7). Their automa-

TABLE 6.7

Maintenance Needs Identified by Kansas City Power & Light's
Capacitor Automation System Based on Two Years of Data

Problem	Annual Percent Failures
Primary fuse to capacitor blown (nuisance fuse operation)	9.1
Failed oil switches	8.1
Hardware accidentally set to "Local" or "Manual"	4.2
Defective capacitor unit	3.5
Miscellaneous	2.4
Control power transformer	1.5
TOTAL	28.8

Source: Goeckeler, C., "Progressive Capacitor Automation Yields Economic and Practical Benefits at KCPL," *Utility Automation*, October 1999.

tion with two-way communications allowed them to readily identify bank failures. The failure rates in Table 6.7 are high, much higher than most distribution equipment. Capacitor banks are complicated; they have a lot of equipment to fail; yet, failure rates should be significantly better than this.

An EPRI survey on capacitor reliability found wide differences in utilities' experience with capacitors (EPRI 1001691, 2002). Roughly one-third of survey responses found feeder capacitors "very good," another one-third found them "typical of line equipment," and the final third found them "problematic." The survey along with follow-up contacts highlighted several issues:

- *Misoperation of capacitor fuses* — Many utilities have operations of fuses where the capacitor bank is unharmed. This can unbalance circuit voltages and reduce the number of capacitors available for var support. Review fusing practices to reduce this problem.

- *Controllers* — Controllers were found "problematic" by a significant number of utilities. Some utilities had problems with switches and with the controllers themselves.

- *Lightning and faults* — In high-lightning areas, controllers can fail from lightning. Controllers are quite exposed to lightning and power-supply overvoltages during faults. Review surge protection practices and powering and grounding of controllers.

- *Human element* — Many controllers are set up incorrectly. Some controllers are hard to program. And, field crews often do not have the skills or proper attitudes toward capacitors and their controls. At some utilities, crews often manually switch off nearby capacitors (and often forget to turn them back on after finishing their work). To reduce these problems, properly train crews and drive home the need to have capacitors available when needed.

6.9   Failure Modes and Case Ruptures

Capacitors can fail in two modes:

- *Low current, progressive failure* — The dielectric fails in one of the elements within the capacitor (see Figure 6.11). With one element shorted, the remaining elements in the series string have increased voltage and higher current (because the total capacitive impedance is lower). With more stress, another element may short out. Failures can cascade until the whole string shorts out. In this scenario, the current builds up slowly as elements successively fail.

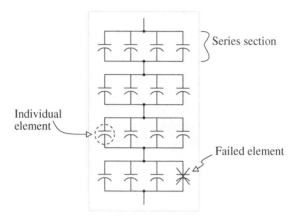

FIGURE 6.11
Capacitor unit with a failed element.

- *High current* — A low-impedance failure develops across the capacitor terminals or from a phase terminal to ground. A broken connector could cause such a fault.

Most failures are progressive. Sudden jumps to high current are rare. To detect progressive failures quickly, fusing must be very sensitive. Film-foil capacitors have few case ruptures — much less than older paper units. An EPRI survey of utilities (EPRI 1001691, 2002) found that film-foil capacitor ruptures were rare to nonexistent. This contrasts sharply with paper capacitors, where Newcomb (1980) reported that film/paper capacitors ruptured in 25% of failures.

Paper and paper-film capacitors have an insulating layer of paper between sheets of foil. When a breakdown in a pack occurs, the arc burns the paper and generates gas. In progressive failures, even though the current is only somewhat higher than normal load current, the sustained arcing can create enough gas to rupture the enclosure. Before 1975, capacitors predominantly used polychlorinated biphenyls (PCB) as the insulating liquid. Environmental regulations on PCB greatly increased the costs of cleanup if these units ruptured (U.S. Environmental Protection Agency 40 CFR Part 761 Polychlorinated Biphenyls (PCBs) Manufacturing, Processing, Distribution in Commerce, and Use Prohibitions). The environmental issues and safety concerns led utilities to tighten up capacitor fusing.

In modern film-foil capacitors, sheets of polypropylene film dielectric separate layers of aluminum foil. When the dielectric breaks down, the heat from the arc melts the film; the film draws back; and the aluminum sheets weld together. With a solid weld, a single element can fail and not create any gas (the current is still relatively low). In film-foil capacitors, the progressive failure mode is much less likely to rupture the case. When all of the packs in series fail, high current flows through the capacitor. This can generate enough heat and gas to rupture the capacitor if it is not cleared quickly.

Figure 6.12 shows capacitor-rupture curves from several sources. Most case-rupture curves are based on tests of prefailed capacitors. The capacitors are failed by applying excessive voltage until the whole capacitor is broken down. The failed capacitor is then subjected to a high-current short-circuit

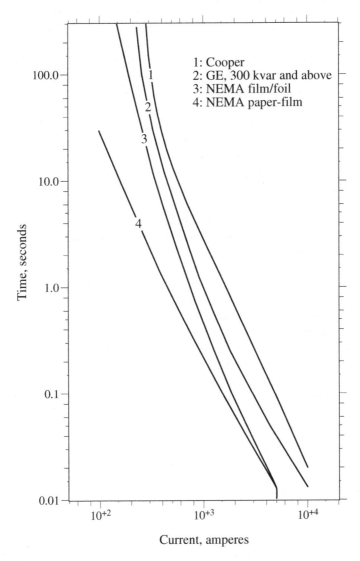

FIGURE 6.12
Capacitor rupture curves. (Data from [ANSI/IEEE Std. 18-1992; Cooper Power Systems, 1990; General Electric, 2001].)

source of known amperage for a given time. Several such samples are tested to develop a case-rupture curve.

The case-rupture curves do not represent all failure modes. Such curves do not show the performance during the most common failures: low-current and progressive element failures (before all elements are punctured). Although, thankfully, rare, high-current faults more severe than those tested for the rupture curves are possible. An arc through the insulating dielectric fluid can generate considerable pressure. Pratt et al. (1977) performed tests on film/foil capacitor units with arc lengths up to 3 in. (7.6 cm) in length. They chose 3 in. as the maximum realistic arc length in a capacitor as the gap spacing between internal series section terminals. Under these conditions, they damaged or ruptured several units for currents and times well below the capacitor rupture curves in Figure 6.12.

Also consider other equipment at a capacitor bank installation. Capacitor switches, especially oil switches, are vulnerable to violent failure. This type of failure has not received nearly the attention that capacitor ruptures or distribution transformer failures have. Potential transformers, current transformers, controller power-supply transformers, and arresters: these too can fail violently. Any failure in which an arc develops inside a small enclosure can rupture or explode. In areas with high fault current, consider applying current-limiting fuses. These will help protect against violent failures of capacitor units, switches, and other accessories in areas with high fault current.

When one element fails and shorts out, the other series sections have higher voltage, and they draw more current. Capacitor packs are designed with a polypropylene film layer less than one mil thick (0.001 in. or 0.025 mm), which is designed to hold a voltage of 2000 V. Table 6.8 shows the number of series sections for several capacitors as reported by Thomas (1990). More recent designs could have even fewer groups. One manufacturer uses three series sections for 7.2 to 7.96 kV units and six series sections for 12.47 to 14.4 kV units. As series sections fail, the remaining elements must hold increasing voltage, and the capacitor draws more current in the same proportion. Figure

TABLE 6.8

Number of Series Sections in Different Voltage Ratings

Unit Voltage, V	Manufacturer		
	A	B	C
2,400	2	2	2
7,200	4	4	4
7,620	5	5	4
13,280	8	8	7
13,800	8	8	—
14,400	8	8	8

Source: Thomas, E. S., "Determination of Neutral Trip Settings for Distribution Capacitor Banks," IEEE Rural Electric Power Conference, 1990. With permission. ©1990 IEEE.

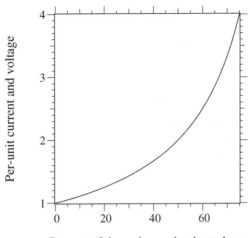

Percent of the series packs shorted out

FIGURE 6.13
Per-unit current drawn by a failing bank depending on the portion of the bank that is failed (assuming an infinite bus). This is also the per-unit voltage applied on the series sections still remaining.

6.13 shows the effect on the per-unit current drawn by a failing unit and the per-unit voltage on the remaining series sections.

If a capacitor bank has multiple units on one phase and all units are protected by one fuse (group fusing), the total bank current should be considered. Consider a bank with two capacitor units. If one unit loses half of its series sections, that unit will draw twice its nominal current. The group — the two units together — will draw 1.5 times the nominal bank load. (This is the current that the fuse sees.)

6.10 Fusing and Protection

The main purpose of the fuse on a capacitor bank is to clear a fault if a capacitor unit or any of the accessories fail. The fuse must clear the fault quickly to prevent any of the equipment from failing violently. Ruptures of capacitors have historically been problematic, so fusing is normally tight. Fuses must be sized to withstand normal currents, including harmonics.

A significant number of utilities have problems with nuisance fuse operations on capacitor banks. A fuse is blown, but the capacitors themselves are still functional. These blown fuses may stay on the system for quite some time before they are noticed (see Figure 6.14). Capacitors with blown fuses increase voltage unbalance, can increase stray voltages, and increase losses. Even if the capacitor controller identifies blown fuses, replacement adds extra maintenance that crews must do.

FIGURE 6.14
Capacitor bank with a blown fuse. (Copyright © 2002. Electric Power Research Institute. 1001691. *Improved Reliability of Switched Capacitor Banks and Capacitor Technology.* Reprinted with permission.)

IEEE guides suggest selecting a fuse capable of handling 1.25 to 1.35 times the nominal capacitor current (IEEE Std. C37.48-1997); a 1.35 factor is most common. Three factors can contribute to higher than expected current:

- *Overvoltage* — Capacitive current increases linearly with voltage, and the reactive vars increase as the square of the voltage. When estimating maximum currents, an upper voltage limit of 110% is normally assumed.

- *Harmonics* — Capacitors can act as a sink for harmonics. This can increase the peak and the rms of the current through the capacitor. Additionally, grounded three-phase banks absorb zero-sequence harmonics from the system.

- *Capacitor tolerance* — Capacitors were allowed to have a tolerance to +15% above their rating (which would increase the current by 15%).

Most fusing practices are based on fusing as tightly as possible to prevent case rupture. So, the overload capability of fuse links is included in fuse sizing. This effectively allows a tighter fusing ratio. K and T tin links can be overloaded to 150%, so for these links with a 1.35 safety factor, the smallest size fuse that can be used is

$$I_{min} = \frac{1.35I_1}{1.5} = 0.9I_1$$

where
I_{min} = minimum fuse rating, A
I_1 = capacitor bank current, A

Table 6.9 shows one manufacturer's recommendations based on this tight-fusing approach.

With this tight-fusing strategy, fuses must be used consistently. If silver links are used instead of tin links, the silver fuses can blow from expected levels of current because silver links have no overload capability.

Prior to the 1970s, a fusing factor of 1.65 was more common. Due to concerns about case ruptures and PCBs, the industry went to tighter fusing factors, 1.35 being the most common. Because of the good performance of all-film capacitors and problems with nuisance fuse operations, consider a

TABLE 6.9

Fusing Recommendations for ANSI Tin Links from One Manufacturer

3-Phase Bank kvar	System Line-to-Line Voltage, kV							
	4.2	4.8	12.5	13.2	13.8	22.9	24.9	34.5
Recommended Fuse Link								
150	20T	20T	8T	6T	6T			
300	40K	40K	15T	12T	12T	8T	8T	5T
450	65K	50K	20T	20T	20T	10T	10T	8T
600	80K	65K	25T	25T	25T	15T	15T	10T
900		100K	40K	40K	40K	20T	20T	15T
1200			50K	50K	50K	30T	25T	20T
1800			80K	80K	80K	40K	40K	30K
2400			100K	100K	100K	65K	50K	40K
Fusing Ratio for the Recommended Link (Link Rating/Nominal Current)								
150	0.96	1.11	1.15	0.91	0.96			
300	0.96	1.11	1.08	0.91	0.96	1.06	1.15	1.00
450	1.04	0.92	0.96	1.02	1.06	0.88	0.96	1.06
600	0.96	0.90	0.90	0.95	1.00	0.99	1.08	1.00
900		0.92	0.96	1.02	1.06	0.88	0.96	1.00
1200			0.90	0.95	1.00	0.99	0.90	1.00
1800			0.96	1.02	1.06	0.88	0.96	1.00
2400			0.90	0.95	1.00	1.07	0.90	1.00

Note: This is not the manufacturer's most up-to-date fusing recommendation. It is provided mainly as an example of a commonly applied fusing criteria for capacitors.

Source: Cooper Power Systems, *Electrical Distribution — System Protection*, 3rd ed., 1990.

looser fusing factor, possibly returning to the 1.65 factor. Slower fuses should also have fewer nuisance fuse operations.

Capacitors are rated to withstand 180% of rated rms current, including fundamental and harmonic currents. Fusing is normally not based on this limit, and is normally much tighter than this, usually from 125 to 165% of rated rms current. Occasionally, fuses in excess of 180% are used. In severe harmonic environments (usually in commercial or industrial applications), normally fuses blow before capacitors fail, but sometimes capacitors fail before the fuse operates. This depends on the fusing strategy.

If a capacitor bank has a blown fuse, crews should test the capacitors before re-fusing. A handheld digital capacitance meter is the most common approach and is accurate. Good multimeters also can measure a capacitance high enough to measure the capacitance on medium-voltage units. There is a chance that capacitance-testers may miss some internal failures requiring high voltage to break down the insulation at the failure. Measuring the capacitance on all three phases helps identify units that may have partial failures. Partial failures show up as a change in capacitance. In a partial failure, one of several series capacitor packs short out; the remaining packs appear as a lower impedance (higher capacitance). As with any equipment about to be energized, crews should visually check the condition of the capacitor unit and make sure there are no bulges, burn marks, or other signs that the unit may have suffered damage.

Some utilities have problems with nuisance fuse operations on distribution transformers. Some of the causes of capacitor fuse operations could be the same as transformer fuse operations, but some differences are apparent:

- Capacitor fuses see almost continuous full load (when the capacitor is switched in).
- Capacitor fuses tend to be bigger. The most common transformer sizes are 25 and 50 kVA, usually with less than a 15 A fuse. Typical capacitor sizes are 300 to 1200 kvar with 15 to 65 A fuses.
- Both have inrush; a capacitor's is quicker.
- Transformers have secondary faults and core saturation that can contribute to nuisance fuse operations; capacitors have neither.

Some possible causes of nuisance fuse operations are

- *Lightning* — Capacitors are a low impedance to the high-frequency lightning surge, so they naturally attract lightning current, which can blow the fuse. Smaller, faster fuses are most prone to lightning. Given that the standard rule of thumb that a fuse at least as big as a 20K or a 15T should prevent nuisance operations, it is hard to see how lightning itself could cause a significant number of fuse operations (as most capacitor bank fuses are larger than this).

- *Outrush to nearby faults* — If a capacitor dumps its stored charge into a nearby fault, the fuse can blow. Capacitor banks also have inrush every time they are switched in, but this is well below the melt point of the fuse.

- *Severe harmonics* — Harmonics increase the current through the fuse.

- *Animal or other bushing faults* — A fault across a bushing due to an animal, contamination on the bushing, or tree contact can blow a fuse. By the time anyone notices the blown fuse, the squirrel or branch has disappeared. Use animal guards and covered jumpers to reduce these.

- *Mechanical damage and deterioration* — Corrosion and vibration can weaken fuse links. On fuse links collected from the field on transformers, Ontario Hydro found that 3% had broken strain wires (CEA 288 D 747, 1998). Another 15% had braids that were brittle and had broken strands. Larger fuses used in capacitors should not have as much of a problem.

- *Installation errors* — Fuses are more likely to blow if crews put in the wrong size fuse or wrong type fuse or do not properly tighten the braid on the fuse.

Outrush is highlighted as a possible failure mode that has been neglected by the industry. Outrush is sometimes considered for station banks to calculate the probability of a fuse operation from a failure of an adjacent parallel unit. But for distribution fuses, nearby faults have not been considered in regard to the effects on fuse operations.

The energy input into the fuse during outrush depends on the line resistance between the capacitor and the fault (see Figure 6.15). The capacitor has stored energy; when the fault occurs, the capacitor discharges its energy into the resistance between the capacitor and the fault. Closer faults cause more energy to go into the fuse. The I^2t that the fuse suffers during outrush to a line-to-ground fault is

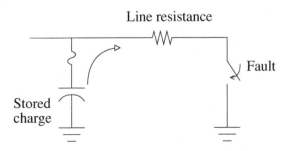

FIGURE 6.15
Outrush from a capacitor to a nearby fault.

$$I^2 t = \frac{\frac{1}{2}CV_{pk}^2}{R} = \frac{2.65Q_{kvar}}{R}V_{pu}^2$$

where

C = capacitance of one unit, µF

V_{pk} = peak voltage on the capacitor at the instant of the fault, kV

R = resistance between the capacitor and the fault, Ω

Q_{kvar} = single-phase reactive power, kvar

V_{pu} = voltage at the instant of the fault in per unit of the capacitor's rated voltage

Table 6.10 shows several sources of fuse operations and the $I^2 t$ that they generate for a 900-kvar bank at 12.47 kV. The nominal load current is 41.7 A. Utilities commonly use 40 or 50-A fuses for this bank. The table shows the minimum melt $I^2 t$ of common fuses. Outrush to nearby faults produces high enough energy to blow common fuses, especially the K links. Of the other possible causes of fuse operation, none are particularly high except for a lightning first stroke. The lightning data is misleading because much of the first stroke will go elsewhere — usually, the line flashes over, and much of the lightning current diverts to the fault.

Use Figure 6.16 to find outrush $I^2 t$ for other cases. Two factors make outrush worse:

- *Higher system voltages* — The outrush $I^2 t$ stays the same with increases in voltage for the same size capacitor bank. The line impedance stays the same for different voltages. But higher-voltage capac-

TABLE 6.10

Comparison of $I^2 t$ of Events that Might Blow a Fuse to the Capability of Common Fuses for a Three-Phase, 900-kvar Bank at 12.47 kV (I_{load} = 41.7 A)

Source	$I^2 t$, A²-sec
Lightning, median 1st stroke	57,000
Lightning, median subsequent stroke	5,500
Inrush at nominal voltage (I_{SC}=5 kA, X/R=8)	4,455
Inrush at 105% voltage	4,911
Outrush to a fault 500-ft away (500-kcmil AAC)	20,280
Outrush to a fault 250-ft away (500-kcmil AAC)	40,560
Outrush to a fault 250-ft away with an arc restrike[a]	162,240
40K fuse, minimum melt $I^2 t$	36,200
50K fuse, minimum melt $I^2 t$	58,700
40T fuse, minimum melt $I^2 t$	107,000

[a] Assumes that the arc transient leaves a voltage of 2 per unit on the capacitor before the arc restrikes.

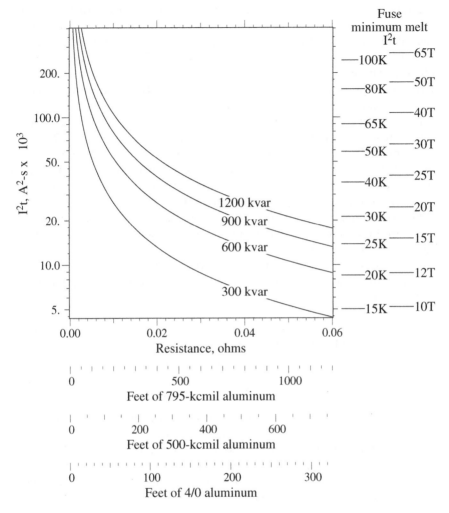

FIGURE 6.16
Outrush as a function of the resistance to the fault for various size capacitor banks (the sizes given are three-phase kvar; the resistance is the resistance around the loop, out and back; the distances are to the fault).

itor banks use smaller fuses, with less I^2t capability. So, a 25-kV capacitor installation is more likely to have nuisance fuse operations than a 12.5-kV system.

- *Larger conductors* — Lower resistance.

Consider a 1200-kvar bank with 500-kcmil conductors. At 12.47 kV (I_{load} = 55.6 A) with a 65K fuse, the fuse exceeds its minimum melt I^2t for faults up to 150 ft away. At 24.94 kV (I_{load} = 27.8 A) with a 30K fuse, the fuse may melt for faults up to 650 ft away. At 34.5 kV (I_{load} = 20.1 A) with a 25 K fuse, the location is off of the chart (it is about 950 ft). Note that the distance scales in Figure 6.16 do not include two important resistances: the capacitor's internal resistance and the fuse's resistance. Both will help reduce the I^2t. Also, the minimum melt I^2t values of the fuses in Figure 6.16 are the 60-Hz values. For high-frequency currents like an outrush discharge, the minimum melt I^2t of expulsion fuses is 30 to 70% of the 60-Hz I^2t (Burrage, 1981).

As an estimate of how much outrush contributes to nuisance fuse operations, consider a 900-kvar bank at 12.47 kV with 40K fuses. We will estimate that the fuse may blow or be severely damaged for faults within 250 ft (76 m). Using a typical fault rate on distribution lines of 90 faults/100 mi/year (56 faults/100 km/year), faults within 250 ft (75 m) of a capacitor occur at the rate of 0.085 per year. This translates into 8.5% fuse operations per capacitor bank per year, a substantial number.

The stored energy on the fault depends on the timing of the fault relative to the point on the voltage wave. Unfortunately, most faults occur at or near the peak of the sinusoid.

Several system scenarios could make individual instances worse; most are situations that leave more than normal voltage on the capacitor before it discharges into the fault:

- *Regulation overvoltages* — Voltages above nominal increase the outrush energy by the voltage squared.

- *Voltage swells* — If a line-to-ground fault on one phase causes a voltage swell on another and the fault jumps to the "swelled" phase, higher-than-normal outrush flows through the fuse.

- *Arc restrikes* — If a nearby arc is not solid but sputters, arc restrikes, much like restrikes of switches, can impress more voltage on the capacitor and subject the fuse to more energy, possibly much larger voltage depending on the severity. (I know of no evidence that this occurs regularly; most arcs are solid, and the system stays faulted once the arc bridges the gap.)

- *Lightning* — A nearby lightning strike to the line can charge up the capacitor (and start the fuse heating). In most cases, the lightning will cause a nearby flashover, and the capacitor's charge will dump right back through the fuse.

- *Multiple-phase faults* — Line-to-line and three-phase faults are more severe for two reasons: the voltage is higher, and the resistance is lower. For example, on a line-to-line fault, the voltage is the line-to-line voltage, and the resistance is the resistance of the phase wires (rather than the resistance of a phase wire and the neutral in series).

These estimates are conservative in that they do not consider skin effects, which have considerable effect at high frequencies. Skin effects increase the conductor's resistance. The transients oscillate in the single-digit kilohertz range. At these frequencies, conductor resistance increases by a factor of two to three. On the negative side, the fuse element is impacted by skin effects, too — higher frequency transients cause the fuse to melt more quickly.

Capacitors also have inrush every time they are energized. Inrush into grounded banks has a peak current (IEEE Std. 1036-1992) of

$$I_{pk} = 1.41\sqrt{I_{SC}I_1}$$

where
I_{pk} = peak value of inrush current, A
I_{SC} = available three-phase fault current, A
I_1 = capacitor bank current, A

The energy into a fuse from inrush is normally very small. It subjects the capacitor fuse to an I^2t (in A^2-s) (Brown, 1979) of

$$I^2t = 2.65\sqrt{1 + k^2}\,I_{SC}I_1 / 1000$$

where
k = X/R ratio at the bank location

Inrush is much worse if a capacitor is switching into a system with a nearby capacitor. The outrush from the already-energized bank dumps into the capacitor coming on line. Fuses at both banks see this transient. In substation applications, this *back-to-back* switching is a major design consideration, often requiring insertion of reactors between banks. For distribution feeder capacitors, the design constraints are not as large. A few hundred feet of separation is enough to prevent inrush/outrush problems. For back-to-back switching, the I^2t is almost the same as that for outrush:

$$I^2t = \frac{\frac{1}{2}CV_{pk}^2}{R} = \frac{2.65Q_{kvar}}{R}V_{pu}^2$$

The only difference is that the capacitance is the series combination of the two capacitances: $C=C_1C_2/(C_1+C_2)$, and $Q_{kvar}=Q_1Q_2/(Q_1+Q_2)$. For the same size banks, $C=C_1/2$, and $Q_{kvar}=Q_1/2$. Figure 6.16 applies if we double the kvar values on the curves. In most situations, maintaining a separation of 500 ft between capacitor banks prevents fuse operations from this inrush/outrush. Separate capacitor banks by 500 ft (150 m) on 15-kV class circuits to avoid inrush problems. Large capacitor banks on higher voltage distribution systems may require modestly larger separations.

Preventing case ruptures is a primary goal of fusing. The fuse should clear before capacitor cases fail. Figure 6.17 shows capacitor rupture curves compared against fuse clearing curves. The graph shows that there is considerable margin between fuse curves and rupture curves. Consider a 12.47-kV, 900-kvar bank of three 300-kvar units, which has a nominal current of 41.7 A. Utilities commonly use a 40 or 50 K fuse for this bank. Larger fuses for this bank are possible, while still maintaining levels below case rupture curves. An EPRI survey found that case ruptures on modern film-foil capacitors are rare (EPRI 1001691, 2002). This gives us confidence that we can loosen fusing practices without having rupture problems.

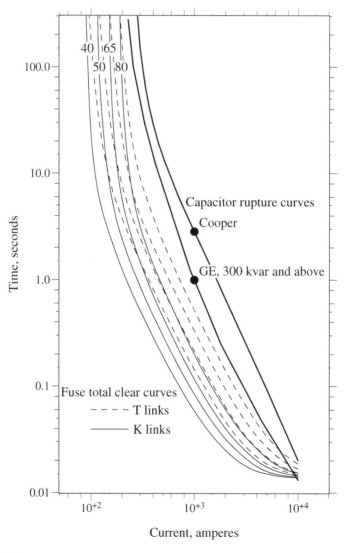

FIGURE 6.17
Fuse curves with capacitor rupture curves.

In areas of high fault current, current-limiting fuses provide extra safety. Either a backup current-limiting fuse in series with an expulsion link or a full-range current-limiting fuse is an appropriate protection scheme in high fault-current areas. While it may seem that expulsion fuses provide adequate protection even to 8 kA (depending on which rupture curve we use), current-limiting fuses provide protection for those less frequent faults with longer internal arcs. They also provide protection against failures in the capacitor switches and other capacitor-bank accessories. Utilities that apply current-limiting fuses on capacitors normally do so for areas with fault currents above 3 to 5 kA.

With backup current-limiting fuses, it is important that crews check the backup fuse whenever the expulsion link operates. On transformers, crews can get away with replacing the expulsion link. If the transformer still does not have voltage, they will quickly know that they have to replace the backup link. But, on capacitors, there is no quick indication that the backup-fuse has operated. Crews must check the voltage on the cutout to see if the backup fuse is operational; or crews should check the capacitor neutral current after replacing the expulsion link to make sure it is close to zero (if all three phases are operational, the balanced currents cancel in the neutral). In addition to not fixing the problem, failing to replace a blown backup fuse could cause future problems. The backup fuse is not designed to hold system voltage continuously — they are not an insulator. Eventually, they will track and arc over.

Because of utility problems with nuisance fuse operations, some loosening of fusing practices is in order. For most of the possible causes of nuisance-fuse operations, increasing the fuse size will decrease the number of false operations. Going to a slower fuse, especially, helps with outrush and other fast transients. If you have nuisance fuse operation problems, consider using T links and/or increase the fuse size one or two sizes. Treat these recommendations as tentative; as of this writing, these fusing issues are the subject of ongoing EPRI research, which should provide more definitive recommendations.

Neutral monitoring (Figure 6.18) is another protection feature that some capacitor controllers offer. Neutral monitoring can detect several problems:

- *Blown fuse* — When one capacitor fuse blows, the neutral current jumps to a value equal to the phase current.
- *Failing capacitor unit* — As a capacitor fails, internal groups of series packs short out. Prior to complete failure, the unit will draw more current than normal. Figure 6.19 shows how the neutral current changes when a certain portion of the capacitor shorts out. Capacitors rated from 7.2 to 7.96 kV normally have three or four series sections, so failure of one element causes neutral currents of 25% (for four in series) or 34% (for three in series) of the phase current.

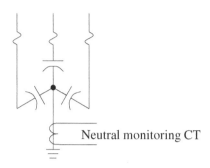

FIGURE 6.18
Neutral monitoring of a capacitor bank.

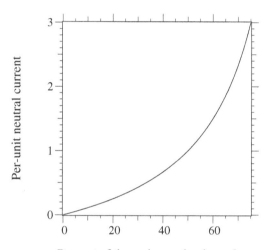

Percent of the series packs shorted out

FIGURE 6.19
Neutral current drawn by a failing grounded-wye bank depending on the portion of the bank that is failed (the neutral current is in per-unit of the nominal capacitor current).

Failure of more than half of the series sections causes more than the capacitor's rated current in the neutral.

- *High harmonic current* — Excessive neutral current may also indicate high harmonic currents.

Neutral monitoring is common in substation banks, and many controllers for switched pole-mounted banks have neutral-monitoring capability. Neutral-current monitors for fixed banks are also available, either with a local warning light or a wireless link to a centralized location.

Neutral monitoring can help reduce operations and maintenance by eliminating regular capacitor patrols and field checks. Quicker replacement of blown fuses also reduces the time that excessive unbalance is present (and the extra losses and possibility of stray voltage). This can lead to more reliable var regulation, and even reduce the number of capacitor banks needed.

6.11 Grounding

Three-phase capacitors can be grounded in a wye configuration or ungrounded, either in a floating-wye or a delta. For multigrounded distribution systems, a grounded-wye capacitor bank offers advantages and disadvantages:

- *Unit failure and fault current* — If a unit fails, the faulted phase draws full fault current. This allows the fuse to blow quickly, but requires fuses to be rated for the full fault current.
- *Harmonics* — The grounded-wye bank can attract zero-sequence harmonics (balanced 3rd, 9th, 15th, ...). This problem is often found in telephone interference cases.

Advantages and disadvantages of the floating-wye, ungrounded banks include

- *Unit failure* — The collapse of voltage across a failed unit pulls the floating neutral to phase voltage. Now, the neutral shift stresses the remaining capacitors with line-to-line voltage, 173% of the capacitor's rating.
- *Fault current* — When one unit fails, the circuit does not draw full fault current — it is a high-impedance fault. This is an advantage in some capacitor applications.
- *Harmonics* — Less chance of harmonic problems because the ungrounded, zero-sequence harmonics (balanced 3rd, 9th, 15th, ...) cannot flow to ground through the capacitor.

The response of the floating-wye configuration deserves more analysis. During a progressive failure, when one series section shorts out, the shift of the neutral relieves the voltage stress on the remaining series sections. In the example in Figure 6.20, for a floating-wye bank with half of the series sections shorted, the line-to-neutral voltage becomes 0.75 per unit. The remaining elements normally see 50% of the line-to-neutral voltage, but now they see 75% (1.5 per unit, so the current is also 1.5 times normal). The reduction in voltage stress due to the neutral shift prolongs the failure — not what we want. The excess heating at the failure point increases the risk of gas generation and case rupture. When one element fails, we really want the fuse (or other protection) to trip quickly. The neutral shift also increases the voltage stress on the units on the other phases.

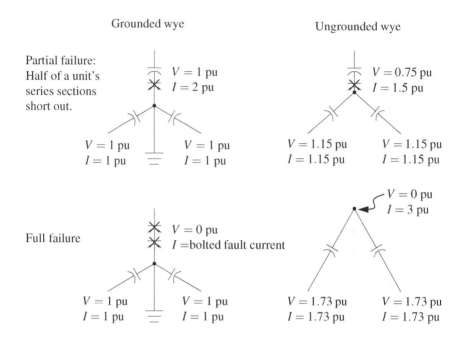

FIGURE 6.20
Comparison of grounded-wye and ungrounded-wye banks during a partial and full failure of one unit. (Copyright © 2002. Electric Power Research Institute. 1001691. *Improved Reliability of Switched Capacitor Banks and Capacitor Technology*. Reprinted with permission.)

Floating-wye configurations are best applied with neutral detection — a potential transformer measuring voltage between the floating neutral and ground can detect a failure of one unit. When one unit fails, a relay monitoring the neutral PT should trip the capacitor's oil or vacuum switch (obviously, this only works on switched banks).

Standard utility practice is to ground banks on multigrounded systems. Over 80% of the respondents to an EPRI survey used grounded-wye capacitor connections (EPRI 1001691, 2002). On three-wire systems, utilities use both ungrounded-wye and delta configurations.

Most utilities use two-bushing capacitors, even though most also use a grounded neutral. Having two bushings allows crews to convert capacitor banks to a floating neutral configuration if telephone interference is a problem.

Utilities universally ground capacitor cases on pole-mounted capacitors (even though it is not strictly required by the National Electrical Safety Code [IEEE C2-1997]). In rare cases, banks with single-bushing capacitors are floated when it becomes necessary to convert a bank to a floating-wye. Avoid this if possible.

References

ANSI C57.12.28-1998, *Pad-Mounted Equipment Enclosure Integrity.*

ANSI/IEEE Std. 18-1992, IEEE *Standard for Shunt Power Capacitors.*

Brown, R. A., "Capacitor Fusing," IEEE/PES Transmission and Distribution Conference and Expo, April 1–6, 1979.

Burrage, L. M., "High Frequency Characteristics of Capacitors and Fuses — Applied in High-Voltage Shunt Banks," IEEE/PES Transmission and Distribution Conference, 1981.

Carlisle, J. C. and El-Keib, A. A., "A Graph Search Algorithm for Optimal Placement of Fixed and Switched Capacitors on Radial Distribution Systems," *IEEE Transactions on Power Delivery,* vol. 15, no. 1, pp. 423–8, January 2000.

CEA 288 D 747, *Application Guide for Distribution Fusing,* Canadian Electrical Association, 1998.

Clark, G. L., "Development of the Switched Capacitor Bank Controller for Independent Phase Switching on the Electric Distribution System," Distributech 2001, San Diego, CA, 2001.

Cooper Power Systems, *Electrical Distribution — System Protection,* 3rd ed., 1990.

DNP Users Group, "A DNP3 Protocol Primer," 2000. www.dnp.org.

EPRI 1001691, *Improved Reliability of Switched Capacitor Banks and Capacitor Technology,* Electric Power Research Institute, Palo Alto, CA, 2002.

Fisher Pierce, "POWERFLEX 4400 and 4500 Series Instruction Manual," 2000.

General Electric, "Case Rupture Curves," 2001. Downloaded from http://www.geindustrial.com.

Goeckeler, C., "Progressive Capacitor Automation Yields Economic and Practical Benefits at KCPL," *Utility Automation,* October 1999.

Grainger, J. J. and Civanlar, S., "Volt/VAr Control on Distribution Systems with Lateral Branches Using Shunt Capacitors and Voltage Regulators. I. The overall problem," *IEEE Transactions on Power Apparatus and Systems,* vol. PAS-104, no. 11, pp. 3278–83, November 1985.

Grainger, J. J. and Lee, S. H., "Optimum Size and Location of Shunt Capacitors for Reduction of Losses on Distribution Feeders," *IEEE Transactions on Power Apparatus and Systems,* vol. PAS-100, no. 3, pp. 1005–18, March 1981.

IEEE C2-1997, *National Electrical Safety Code.*

IEEE Std. 18-2002, *IEEE Standard for Shunt Power Capacitors.*

IEEE Std. 1036-1992, *IEEE Guide for Application of Shunt Power Capacitors.*

IEEE Std. C37.48-1997, *IEEE Guide for the Application, Operation, and Maintenance of High-Voltage Fuses, Distribution Enclosed Single-pole Air Switches, Fuse Disconnecting Switches, and Accessories.*

Lee, S. H. and Grainger, J. J., "Optimum Placement of Fixed and Switched Capacitors on Primary Distribution Feeders," *IEEE Transactions on Power Apparatus and Systems,* vol. PAS-100, no. 1, pp. 345–52, January 1981.

Neagle, N. M. and Samson, D. R., "Loss Reduction from Capacitors Installed on Primary Feeders," *AIEE Transactions,* vol. 75, pp. 950–9, Part III, October 1956.

Newcomb, G. R., "Film/Foil Power Capacitor," *IEEE International Symposium on Electrical Insulation,* Boston, MA, 1980.

Pratt, R. A., Olive, W. W. J., Whitman, B. D., and Brown, R. W., "Capacitor Case Rupture Withstand Capability and Fuse Protection Considerations," EEI T&D Conference, Chicago, IL, May 5-6, 1977.

RUS 1724D-112, *The Application of Capacitors on Rural Electric Systems*, United States Department of Agriculture, Rural Utilities Service, 2001.

Shyh, J. H., "An Immune-Based Optimization Method to Capacitor Placement in a Radial Distribution System," *IEEE Transactions on Power Delivery*, vol. 15, no. 2, pp. 744–9, April 2000.

Thomas, E. S., "Determination of Neutral Trip Settings for Distribution Capacitor Banks," IEEE Rural Electric Power Conference, 1990.

They quit respecting us when we got soft and started using bucket trucks and now anybody can become a lineman, it is sick.

In response to: Do linemen feel they are respected by management and coworkers for the jobs they are doing, do management and coworkers understand what you do?

www.powerlineman.com

Index

311

T - #0036 - 111024 - C0 - 229/152/19 - PB - 9780367391676 - Gloss Lamination